Water – The Last Chance?

Water – The Last Chance?

How to Manage Limited Water Resources

Michael Rouse

IWA PUBLISHING

Published by **IWA Publishing**
Unit 104–105, Export Building
1 Clove Crescent
London E14 2BA, UK
Telephone: +44 (0)20 8054 8202
Email: publications@iwap.co.uk
Web: www.iwaponline.com

First published 2026

British Library Cataloguing in Publication Data
A CIP catalogue record for this book is available from the British Library

ISBN: 9781789065510 (paperback)
ISBN: 9781789065527 (eBook)
ISBN: 9781789065534 (ePub)

Doi: 10.2166/9781789065527

This eBook was made Open Access in October 2026.

Contents

Chapter 7

Chapter 8

Chapter 9

Foreword

Water underpins human existence: our health, our food, the energy that powers economies, and an estimated 1.7 billion jobs across agriculture, industry, and services. But for billions of people, water insecurity persists and is a fundamental barrier to growth. Three and a half billion people lack safely managed sanitation. Two billion lack safe drinking water. Four billion face water scarcity. And under current agricultural practices, 6.5 billion people will not have sustainable food sources by 2050. These are not distant projections – they are the trajectory we are already on.

For decades, global targets on water – from the 1980 Water Decade, to the 2002 Millennium Development Goals and to the Sustainable Development Goals – were set but action fell short.

But the tide can – and must – turn.

We know now much better which solutions work, why we need to capture more rainfall and to reduce water use, and how to overcome bottlenecks such as the absence of country ownership, weak policy fundamentals, and fragmented financing.

This book, based on the accumulation of 50 years of experience across the globe on what works in practice, provides the vehicle to channel political and professional attention to reduce water deprivation and famine. In addition to setting out what needs to be done it very importantly describes how.

Strengthening water security is not solely a technical problem to be solved by experts. It is a political and economic priority that demands action at every level – from national policymakers to local governments, and from system approaches to individual behaviours. Every reader of this book has a role to play. And when we all contribute, we not only halt a crisis, we also unlock the full potential of people and economies in a water-secure future.

Saroj Kumar Jha
World Bank Group Director, Water Knowledge Bank
Co-chair 2030 Water Resources Group

doi: 10.2166/9781789065527_xiii

Preface

THIS BOOK IS ABOUT 'THE HOW'

Shortly after I began drafting this book, in January 2026 the United Nations issued its report titled *Global Water Bankruptcy: Living Beyond our Hydrological Means in the Post-Crisis Era.* This well-researched update on global water, was prepared by the United Nations University Institute for Water, Environment and Health (UNU-INWEH). Its main message is that it is no longer a matter of 'pulling up our socks' but a need to establish a new global water agenda as existing ones and conventional policies are not sufficient for the world's current water reality. At the beginning of chapter 1 on Policies, planning, regulation and communications, I list the key policy messages in that report. Ironically, in 2006 when I was drafting my 2007 book *Institutional Governance and Regulation of Water Services*, UNDP published its Human Development Report 2006 (*Beyond Scarcity – Power, Poverty and the Global Water Crisis*).

Over the years there have been a number of reports identifying 'what' needs to be done, starting from the Water Decade in 1980, which lasted 20 years, the Millennium Development Goals (MDGs), the Sustainable Development Goals (SDGs) and most recently the output from the UN Water Ministers Conference in March 2023, which produced some valuable 'goals' relating to multiple parameters such as demand management for both communities and irrigation, and the need for the development of early warning systems for both floods and droughts. A number of countries made financial commitments, but it appears that many of those have not been substantiated. The March 2023 initiatives are no longer being discussed.

It is interesting to observe that the more recent goals are very similar to those in the Water Decade. The data in the Global Water Bankruptcy report shows that in some respects the world has been going backwards by ignoring the problem, for example, the lack of concern for water storage, particularly aquifer management. Looking back at the preface in my 2007 book, I said that establishing effective regulatory structures was difficult because it involved not just science or engineering, but also people.

I have been conscious all along that the international goals all related to 'the what' and not 'the how'. Over the years I have developed databases which relate to methodology with case studies. My teaching on the Water Management Module of the University of Oxford MSc Course Water Science, Policy and Management addresses 'the how' across community systems, irrigation and water for industry, together with early warning systems. Some of the content of this book is based on that lecture material.

The content of the book will probably never be complete as methodologies evolve and new case studies emerge. Open access will facilitate updates. I hope that water managements, whether related to policy, to regulation and particularly to implementation, will find that the book provides them with 'the how' that they need, and that students will find the book assists them in their studies and stimulates them to take on future water challenges.

Acknowledgements

I am sorry that I cannot list the names of all who have helped to shape my views over the 50 years I have worked in the world of water. I am grateful for their collective wisdom and experience. Those who work in water are dedicated people, passionate about their work and the services they provide, and willing to share their experience for the benefit of all. I thank them all and am grateful for the opportunity and good luck to be part of that international community. I hope that at least some of them will feel that I have done them justice in trying to assemble the collective wisdom on matters relating to water management.

In my international work, during my time as President of the International Water Association and as a consultant, I have been privileged to meet many excellent people involved in water. They have been from the different types of organisations; national government, local government, regulation, public and private utilities, trade associations, manufacturers, international associations, charities and academia. They have different perspectives on situations, and approach issues from different viewpoints and contribute to the establishment of effective policies. In producing a 'how to' book it has been important for me to recognise this as it is necessary for change to meet today's challenges.

It is inevitable that the greatest influence has come from my colleagues, firstly at the Water Research Centre (WRc), then at the Drinking Water Inspectorate for England and Wales, and on to the University of Oxford MSc Course on Water Science, Policy and Management (WSPM). I am not listing names for fear of omitting someone, but I thank them all.

Water institutions provide opportunities for sharing knowledge across water activities, and I have had the privilege of being President of two water institutions. Firstly, the Institute of Water, which is the UK organisation dedicated to the development of water professionals across the UK water industry, and secondly, the International Water Association, covering both water and wastewater worldwide. For the latter, I was involved in Congresses held on all continents and also attendance at many of its members' own events. I learnt about universal issues but also that 'one size does not fit all', with the need for issues to be considered and developed in different organisational and technical aspects.

Taking my experience and producing structured lectures requires rigour and discipline in research and I thank Rachael McDonnell, the main architect of the University of Oxford MSc Course, WSPM, for her vision in bringing an applied researcher and consultant into the teaching team in 2005. I thank Rob Hope, who followed Rachael as the Course Director, for his support and ongoing encouragement. I thank Kevin Grecksch, the current Course Director, for recognising the value of my knowledge and experience in Water Management. I am particularly grateful to my colleague Alex Money for his input into the chapter on Financing by providing relevant water examples of blended finance and for reviewing the text.

There are now over 500 alumni of the WSPM Course. I thank the current students and the alumni for their probing questions and the stimulus they have provided over the years. I hope that those still active in water will value the book and promote its content in meeting the great challenges of today.

I thank IWA Publishing for recognising the importance of the book content. I thank my partner, Helene Horth, for her encouragement throughout my drafting of the book, for her suggestions on improvements and her diligent proof reading.

doi: 10.2166/9781789065527_0001

Chapter 1

Important policy issues

Some bricks in the wall

BACKGROUND TO POLICIES

Recognising the relevance and importance of the update on the global water scene, I give the key findings of the UN Global Water Bankruptcy report January 2026 (Madani, 2026).

- 75% of the world's population live in countries classified as water insecure, which indicates that water-related risks are now systemic rather than marginal.
- The language of crisis – suggesting a temporary emergency followed by a return to normal through mitigation efforts no longer captures what is happening in some parts of the world
- So, the central message is 'The World has entered the era of Global Bankruptcy' with the following messages:
- How to admit solvency where it exists
- How to manage irreversibility honestly
- How to share unavoidable losses fairly
- How to design institutions, development pathways and financial frameworks that prevent further overspending of hydrological capital

The UN report does not include flooding, the management of which is highly connected to managing rainfall and increasing water storage. In most areas in the past the natural catchment provided for control of runoff through woodlands, natural wetlands, flood plains and permeable surfaces. This not only restricted flooding but maintained aquifer capacity through natural recharge. Over time, with growing populations, there has been a demand for timber, for grazing and arable farmland. At the same time the growth in housing, with increased roads and paving and lack of enforcement of building regulations has produced increasing flows upstream of rivers, and mixing storm runoff with sewage has produced dirty and unhealthy rivers and contaminated beaches. Climate change with more intense storms and longer periods of drought has resulted in both greater floods and greater demand by populations, and irrigation with water supplied from storage, particularly aquifers. The current situation is a backlog of investment with longer timetables than the duration of most governments. Managing rainfall and achieving great storage are the two fundamentals irrespective of the current state of water. This is covered in chapter 6.

It can be expected that the impact around the world will be uneven, developed countries in temperate regions are likely to be less affected, but those areas already subject to severe droughts are those with the greater population growth rates. They tend also to be the suppliers in Virtual Water goods (see chapter 4) to the detriment of their own population's water requirements.

The policy items discussed in this chapter are based on the needs of the post-crisis era and how to address the challenges brought by water bankruptcy.

Organisation structures and responsibilities

These are covered under five sub-headings:

(1) Overall government responsibilities
(2) Separation of policy, regulation and delivery
(3) Vertical integration
(4) Water service providers
(5) Investment, cost recovery, tariffs and affordability

Overall government responsibilities

Overall responsibility for policy generally rests with central governments although, in some countries, it is delegated to a regional level. Important policy issues range from decisions on governmental departmental structures to consideration of more detailed aspects such as drinking water quality. Many government functions have an interest in water. Water relates to health, to the environment, to agriculture, to regional development, to housing and to finance, among others. Because of this, there has been a tendency for the responsibilities to be spread across government departments. In the past, perhaps this didn't matter very much, but with the increasing demand for water in a post-crisis era there is a need for an integrated approach to policy and planning. It is highly desirable that just one government department is involved but, if not, there needs to be a designated lead department, or some form of coordinating committee with sufficient status to overrule territorial obstacles to progress.

Then there is the question of regulation. Who decides on water charges? How should the environmental 'costs' of water be reflected in water prices? Who monitors against the required deliverables including managing vested interests which will have to be limited. Who has responsibility for enforcement of standards? Who will issue reports on drinking water quality and environmental standards?

Separation of policy, regulation and delivery

Under pre-crisis conditions, the traditional structure for water services was the municipal model in which central government, in consultation with local government, determined overall policies, and local government was responsible for detailed aspects of policy, for example, policy on cost recovery, and for delivery and monitoring. Often, the municipal utility monitored its own performance. Water charges were set by local government, with little or no transparency. In many cases the water charges were included in local taxation. Consumers didn't know how much the services cost, or whether or not it was an efficient service. Government's role is policy making, and those involved in government are generally most suited to a policy function. Elected local government officials are interested in politics and policy, and local authority staff have been trained as administrators, not as operational managers. There are exceptions but governments, either national or local, are not generally good at managing operations, and with policy responsibilities having to take priority, the direction and energy of the officials is inevitably taken up in local political matters.

Therefore, in order to achieve effective scrutiny of the operations, it is necessary to introduce some independent body to act as a regulator and auditor of published results. There are clear benefits in having separate policy, regulation and delivery. The challenge is how to achieve integration under water bankruptcy conditions. A single regulator overseen by government and community bodies is recommended.

Vertical integration

I believe that under normal conditions the most desirable form of vertical integration is to have all aspects of water supply, plus the equivalent parts of the wastewater system, managed by the same organisation. With water bankruptcy requiring consistent policies vertical integration may be even more important with the need to extend wastewater reuse. The Singapore example of integrated water resources discussed in chapter 3 is a perfect illustration of the value of having such vertical integration. Others who believe that there are benefits from direct competition will disagree but finding ways of harnessing commercial energy will be important in tackling the major challenges to be faced. Integration of all four parts of water supply (water source, treatment, distribution and wastewater reuse) is important, as it provides not only for the important hydraulic and water quality linkages, but also for the vital relationship between costs, revenue collection and customer service. It is particularly important for the integrated management of water treatment, water distribution and customer service. BOT (Build, Operate and Transfer) projects separate treatment from distribution and customer service. A BOT contractor needs to be confident that there will be an income stream for the duration of the contract, before being willing to commit money to the construction of the plant. This has resulted in contracts in which the utility has been committed to paying for a volume of water, quite independently of the demand or of the supply circumstances at its own treatment plants.

There is also the question of drinking water quality monitoring, with the need for the quality of the water from all sources entering the distribution system to be monitored to reduce the risk of contaminated water reaching consumers.

If treatment and distribution are managed separately, monitoring becomes not only a technical requirement, but also a contractual issue. As recommended above, a single regulator overseen by government and community bodies is recommended.

Water service providers

National or state governments need to determine policies and the roles and responsibilities of water service. Each provider will need a manager with direct jurisdiction over its supply area. Each supply boundary will be determined by the supply system and not by size. It will not be possible to have a management structure covering more than one supply system. It is likely that the service providers will be public suppliers. Any private suppliers will have to be closely regulated to focus on the basic service delivery functions.

Investment cost recovery, tariffs and affordability

Significant investment will be required in both rainfall catchments and in water service delivery. In the catchments as much rainfall as possible needs to be captured, both to reduce flood risks and to provide both additional water for supply and storage. This can only be funded out of the public purse. In water service delivery the most critical investment is in reducing water distribution leakage, this being the most available new source of water. The simple calculation is that reducing leakage from 50% to 30% increases water supply by 40%. Investment in leakage management has been neglected in most parts of the world, so the required investment increase is substantial. The sums involved will have to be built into the water service provider operating budgets. Calling it capital investment with all the implications of borrowing must be avoided. This aspect is discussed in chapter 10. It is important to recognise that at the time of writing this book, the world was in political and financial crisis. However, investment in water will have to be given some priority to prevent world famine from getting much worse.

There are important questions about water tariffs and how the financial burden will be distributed across populations, with consideration of provisions for the poor. This is covered in chapter 2 with a discussion on approaches in a pre-crisis era and how the water bankruptcy era might evolve. Inevitably, there will have to be a transition programme.

The importance of continuous 24/7 water supplies

The importance of water distribution system leakage control was referred to in the last section. The benefit calculation is compelling, but it is important to recognise that in many parts of the world in which demand has outstripped supply it has been necessary to ration supplies to a few hours of the day. Unfortunately, the process of unpressurising and repressurising systems creates pressure surges which damage the systems and increases pipe bursts and leakage (Charalambous, n.d.). To recover from this situation requires systems to be zoned so that zones can be refurbished one at a time to become continuous supplies. So, achieving 24/7 is a vital policy decision. This is discussed fully in chapter 3.

Water resource planning

There has been a tendency for water resource planning to be seen as integrated water resource management (IWRM), with emphasis on river basin management and the question of how a river's water is shared in a sustainable way between people living along its length and the allocation between cities and agriculture. However, most progress has been made where there has been a specific need to be addressed. I think this is important because people come together most effectively when working on a project with a defined practical output. It is for this reason that I like 'drinking water safety plans' which bring together all the organisations involved from catchment to tap; I cover this subject in the next section.

Meeting the challenge has changed with the water bankruptcy need for emphasis on rainfall capture and demand management. Chapters 3, 4 and 5 are devoted to demand management across communities, irrigation and water for industry. Chapter 6, 'Urban Water and Urban Water Management' is concerned with rainfall capture and storage.

Water storage

A major failure over recent years across the world has been a lack of attention to water storage. It is required to provide water during droughts and countries planned based on experience on how many years there will be limited rainfall. For example, parts of Australia have typically planned for seven years or more (South East Queensland, Australia). In the aquifer supplying Lodwar Town in Turkana State, Kenya, the University of Nairobi estimated that the aquifer could not be uncharged for more than 5 years. In temperate areas water security has been thought to require only relatively shorter periods of time. That has now been called into question by the Cape Town experience of 2017 (Smedley, 2023). All countries need to review their water storage situation and consider how it can be enhanced. The question is on how.

All forms of storage should be used according to the terrain. Aquifers which are able to be recharged should be given priority. Distributed local reservoirs and ponds are particularly relevant to irrigation. The water used for 'new' storage should not come from existing sources but from captured rainwater. Short-term actions based on political expediency to use groundwater to create irrigation ponds should be avoided as it would make the longer-term situation worse.

Public participation and steps to minimise corruption

Very high investment is required in catchment management and in water demand management. A high degree of transparency and public information is required to ensure vital investment is not lost due to corruption. An associated requirement is that social media and other potentially exploitative sources of information are not used to diminish the importance of the investments. The public participation aspects of water bankruptcy are discussed in chapter 7.

Water for health

World Health Organisation (WHO) guidelines provide the basis for the quality of drinking water. They are prepared and updated by the world's experts on drinking water quality and health. They refer to (the 'what') of drinking water quality. The 'how' is provided by water safety plans for which I worked with WHO in their development and applied them to the regulation of drinking water quality in Ghana in 2004. They were launched at the IWA Congress in Marrakesh, Morocco, later that year. The approach involves consideration of the risks to quality and the steps to be taken to manage those risks. The other part of the how was introduced by WHO (WHO/Unicef/JMP, n.d.) with the concept of safely managed water supplies with the following rungs.

- **Safely managed**
  Drinking water from an improved water source that is accessible on premises, available when needed and free from faecal and priority chemical contamination
- **Basic**
  Drinking water from an improved source, provided collection time is not more than 30 minutes for a roundtrip including queuing
- **Limited**
  Drinking water from an improved source for which collection time exceeds 30 minutes for a roundtrip including queuing
- **Unimproved**
  Drinking water from an unprotected dug well or unprotected spring
- **Surface water**
  Drinking water directly from a river, dam, lake, pond, stream, canal or irrigation canal.

The JMP 2017 Thematic Report on Safely Managed Drinking Water (WHO/Unicef/JMP, 2017) considers the implications of Sustainable Development Goal 6.1 target. By 2030, the aim is to achieve universal and equitable access to safe and affordable drinking water for all, and outlines JMP plans for enhanced monitoring of drinking water during the sustainable development goal period. It is important that these quality measures are incorporated into water bankruptcy planning.

Water as a human right

The extremely important Human Development Report 2006 (UNDP, 2015), produced by UNDP, with its subtitle *Beyond Scarcity – Power, Poverty and the Global Water Crisis*, stresses the need for political leadership on water and sanitation, and the need for the establishment of 'dedicated water and sanitation ministries led by senior cabinet ministers'. This can be consistent with water bankruptcy measures with consideration for all communities.

Training

Managing the provisions for water bankruptcy and the associated transitional programmes will require the availability of staff trained in the required administrative and technical processes. As part of their national plans, national governments should provide for training. Countries with the skills and the training capability should be gearing up to provide such training. This is a critical part of early planning.

The following chapters

In the preface, I refer to the international water goals presented in all the various attempts to promote improved water services worldwide. I reflected that the goals from 1980 onwards are very similar, and they all relate to 'the what' without consideration of 'the how'. This book provides the how across the critical aspects of water supply. The following chapters relate to the how, and also reflect on necessary transitional programmes.

TRANSITION ARRANGEMENTS

Meeting the challenges of water bankruptcy will require major changes. It will be necessary for these changes to be phased in over time. Transitional planning is essential. Each chapter proposes transitional arrangements relevant to the subject matter covered in it. On policy issues, an initial step is for governments to advise the public that changes are necessary and that transitional arrangements will be put in place.

THE APPENDIX

The Appendix provides key aspects of chapters in my book *Institutional Governance and Regulation of Water Services: The Essential Elements, 2nd edition* (2013), which, although valuable in their own right, don't align with the focus of this book.

REFERENCES

Charalambous B. (n.d.). The Hidden Costs of Resorting to Intermittent Supplies. https://hydrocontrolltd.com/wp-content/uploads/2025/07/The-Hidden-Costs-of-Resorting-To-Intermittent-Supplies.pdf

Madani K. (2026). Global Water Bankruptcy: Living Beyond Our Hydrological Means in the Post-Crisis Era. UNU-INWEH Report, United Nations University Institute for Water, Environment and Health (UNU-INWEH), Richmond Hill, Ontario, Canada. https://doi.org/10.53328/INR26KAM001

Smedley T. (2023). The Last Drop: Solving the World's Water Crisis. Picador, Basingstoke, UK. ISBN 9781529058147.

South East Queensland, Australia. Regional Drought Resilience Plan 2019. https://rdrpqld.org.au/region/south_east_queensland/

UNDP. (2015). Water Rights and Wrongs: a summary of the Human Development Report. https://www.undp.org/publications/water-rights-and-wrongs-summary-human-development-report

WHO/Unicef/JMP. (2017). Progress on drinking water, sanitation and hygiene: 2017 update and SDG baselines. https://washdata.org/reports/jmp-2017-wash-households

WHO/Unicef/JMP. (n.d.). Drinking Water Ladder. https://washdata.org/topics/drinking-water

doi: 10.2166/9781789065527_0009

Chapter 2

Planning and costing

Water: not a free for all

It is argued that water is a human right and as such, should be free. At the Kyoto World Water Forum in 2003, protesters campaigning against the involvement of the private sector in the provision of water services carried placards to this effect. They disrupted serious discussions on how to provide water services to the currently underserved – discussions aimed at meeting the water and sanitation millennium development goals. The Minister responsible for water in Swaziland put this argument in perspective, when he was reported as saying words to the effect that 'certainly water is free, providing you wish to take your bucket up the mountain to the source to collect your water, but if you wish me to make provision for the collection, treatment and delivery of safe drinking water to your home, then I have to charge you for the service'. This chapter discusses finance measures for such a service, how necessary costs have to be recovered, and the critical planning process. It also recognises that the post-crisis era discussed in chapter 1 will influence planning and costing.

WHERE NOT TO BE

Downward spirals or what are often called 'vicious circles', should be avoided. Both terms describe the situations well, although unfortunately the terms are used widely in other contexts. I refer to three circles, the consumer circle, the salary circle and the subsidy circle. I have taken them from a presentation by Eero Kontula, Finnida (Kontula, 2004). The circles (Figure 2.1) are best considered together as the problems they describe are interconnected, with the fundamental cause being unsustainable cost recovery arising from low tariffs. Low tariffs without subsidies mean poor service, discontented consumers and even lower income. Low income means poor salaries, low motivation and the

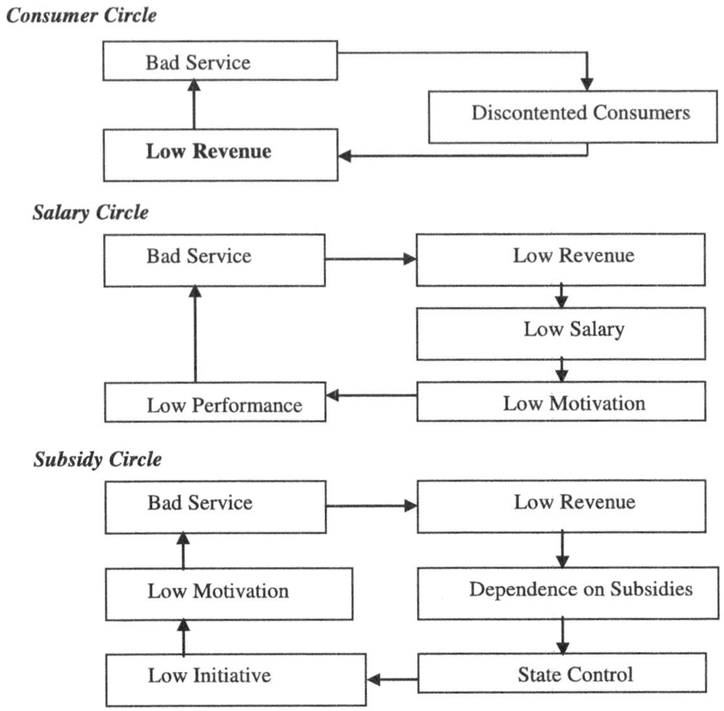

Figure 2.1 The vicious circles of water service decline. (Source: Presentation by Eero Kontula, Finnida, at World Bank/WaterAid Workshop, London, August 2004).

likelihood of corruption, poor performance, discontented consumers and even lower income. Subsidies mean low initiative, low motivation, poor service and, depending on the status of the water service provider, may result in state control.

This situation is not limited to the developing world. In parts of the developed world traditional municipal operations suffer from the same ailments. Probably, water sector workers in developing countries would not see their counterparts in the developed world as being poorly paid, but local government pay is often lower than the private sector. Consequently, local governments can find it difficult to attract high-quality management and staff to work in the water sector. Elected officials, for popularity reasons, are unwilling to raise tariffs, and subsidies are rarely sufficient to achieve sustainable cost recovery (Sustainable cost recovery). The consequence is a downward spiral of deteriorating infrastructure and poor service. Some municipalities have looked to the private sector for salvation but without recognising that the fundamentals of investment in the deteriorating infrastructure, and tariff increases for sustainability, are essential. This is quite independent of whether the systems are managed by the public sector or the private sector. Successful use of the private sector is given in the appendix, reproduced from the chapter in my book *Institutional Governance and Regulation of Water Services*, 2013.

These vicious circles are of even greater consequence in the developing world given the difficulty of raising sufficient funding, and of developing management capacity, as it is more difficult to break out of the downward spirals. There are no quick fixes, it takes time, and there is a need for courageous politicians, either supported or provoked by competent NGOs, to achieve the necessary changes. The increased investment required to deal with water bankruptcy will require greater political courage and public understanding.

I have highlighted where not to be but recognise that, as a result of a lack of attention to important fundamentals, many water service operations are already where they should not be. I now return to the elements of cost.

Donor and bank investment

The first consideration is where is investment most needed? There appears to be a tendency for funders to wish for a recipient country to invest in new treatment plant or enhanced treatment facilities. From a donor's or banks' point of view, it can be well-defined, costed accurately, and it can be delivered to a predictable timetable. Also there remains a 'monument' to an investor's contribution. Local politicians are attracted to these investments in a similar way, especially if the completion can coincide with their period in office, and there is the opportunity of some good publicity. In many cases, additional treatment capacity is the priority, but in many other cases, the priority should be the refurbishment and extension of distribution systems, or a combination of both treatment and distribution capacity. One argument put forward for giving priority to treatment is that long lead times are involved, although in practice the refurbishment of distribution systems takes longer, and is essential in order to deliver increased production capacity.

Also, cities grow not in the best interests of water supply planning, but for many other reasons, including the establishment of unplanned settlements on the outskirts with ongoing movement of people from rural to urban areas. Original distribution systems were designed to supply a defined area with pipes becoming smaller as the branches of the tree network reach the extremities. Therefore, in order to extend systems, it is necessary to include investment in larger pipes, either to reinforce systems in the same overall direction away from treatment works, or to create a feed in from the new extremity back towards the treatment works. The investment required for the latter, and the timetable for construction can exceed that needed for new treatment works. These considerations require integrated planning, which requires sound network analysis, which in turn requires good information on the distribution systems, and sound water demand predictions. Well, the reader may say, this is all very obvious, but many decisions appear to be made without a good understanding of the options and the knock-on implications of any one aspect. The strong recommendation is that investors should begin by providing support for good technical assistance, to assist the operational utilities in learning how to carry out measurements required for accurate water balances, and for calibration of water network analysis models.

Table 2.1 Estimated investment requirement in water and sanitation

Estimated Investment Requirement in Water and Sanitation 2005–2030

Area	US $ 000 000 000 000
Middle East	0.21
Africa	0.23
US/Canada	3.62
South America/Latin America	4.97
Europe	4.53
Asia/Oceania	9.04
TOTAL	22.60

Source: USAID Water and Development Technical Brief 9 (2021).

These tools then provide a firm basis for forward planning on leakage control, extensions to distribution networks and increased resources and treatment. The message here is to start with consumers' needs and work back up the supply system. An excellent example (Ministry for Foreign Affairs of Finland, 2012) of this is the Finnish Government's support to Vietnam.

A USAID Technical Brief paper gives an indication of the costs involved (Table 2.1). Costs are discussed in depth in chapter 10.

The investment needs will be compounded by population growth, estimated to be a further 2.3 billion people by 2050 (World Population, 2026), an increase of 33%, largely in developing countries. With ongoing urbanisation, an estimated additional 2.6 billion urban residents by 2050 (UN DESA, 2017), the challenge is to expand water and sanitation provision in urban areas in developing countries by at least 50% over and above the current deficit. Even without additional population growth it is now recognised that there is insufficient water to support the current population. To quote from the UN Report on Water Bankruptcy published in January 2026 (Madani, 2026) 'The scale of unmet basic needs underscores this gap between rhetoric and reality while 2.2 billion people lack safely managed drinking water and 3.5 billion lack safely managed sanitation.' Even where services exist, they are often fragile. Around 6.1 billion people live in areas that are water-insecure or critically water-insecure. The integration of water and sanitation planning into sustainable city planning will become more and more critical. Planning is discussed later in this chapter.

A vitally important consideration in the case of loans is their cost and payback period, due to the impact on water charges. Water by unit volume is a low-cost product. Capital assets are expensive, and can have a big impact on the cost of water supply, unless the loans have a low interest rate and recovery can be over a period reflecting the life of the asset, say up to 30 years or more. It is for this reason that some concession contracts are for a long period, with cities foregoing the periodic opportunity of competition through re-tendering. Malaysia, recognising the need for long-term low-interest loans, as part of the water sector reforms enacted by the National Water Services Commission Act and Water Services Industry Act 2006 (WIS Act), has formed a government-owned body called Water Asset Management Company (Pengurusan Ast Air

Berhad (PAAB)) to fund and develop the water infrastructure. This company, backed by government security, is able to secure long-term low-interest loans from the international money market. The approach is for PAAB to fund developments approved by the regulator and to recover the costs through leasing charges on the operator. At the end of the leasing period ownership transfers to the operator and that capital becomes available for other developments.

Water charges

For sustainability, water charges should be at a level sufficient to cover fully the costs of operating efficient water services, and to make provision for the necessary refurbishment of the infrastructure. As discussed in the last section, the costs to be recovered may include an element for servicing the capital cost of the infrastructure, depending on whether or not there are grants, subsidies or loans. Determining accurate operating and maintenance costs, and calculating production and distribution unit costs, should be a straightforward process, but often in municipal systems, there have not been separate accounting systems, there is a lack of a cost-management culture, and there is no accurate information on water produced or water delivered to consumers. The need for a good water balance has already been mentioned, but there is also a need to identify actual costs of labour, energy and bought-in items. This then provides a sound basis for considering what the costs should be in a reasonably well-operated system, because consumers should not have to pay for poor procurement practice or for the cost of an inefficient service.

Often, at the beginning of a reform process, charges are not sufficient to cover even the day-to-day operating costs. Sometimes, there are government subsidies to cover part of the operating costs. However, this is poor practice for three reasons. Firstly, subsidies rarely make sufficient provision for infrastructure refurbishment. Secondly, they are unreliable because there are always competing local needs for those monies, with the inevitable consequence that there will be cuts in the amount available for the water service. Thirdly, general subsidies provide greater benefit to the more wealthy consumers, who use more water, than to the poor. Subsidies, where available, should be focussed on the needs of the poor, and this is discussed later in this chapter.

Given good housekeeping, and a plan to improve operational performance over time, there is a need gradually to increase tariffs towards sustainable cost recovery. Consumers are going to be unhappy if tariffs are raised without some tangible improvement in the service, perhaps seen as more hours of water supply per day following some initial effective leakage control. Consumer consultation is particularly important during the transition stage, with explanation of why sustainable cost recovery is essential, and if necessary a plea for some tolerance during the initial stages, as improvements in service may not be seen in the early stages. Indeed, it may be necessary to advise consumers that for a period they will be without a piped supply, and that a tanker service will be provided, while critical parts of the distribution system are being replaced or refurbished. It is important that consumers are given an honest account of the difficulties, and that expectations are not raised above what can be delivered, especially in relation to the time it takes to achieve improvements. It is easy to allow systems to deteriorate, and loss of a good quality service is not immediate, but once systems are in a state of decay, recovery takes time, investment and staff development. How can financial provision be made to prevent deterioration?

Adequate investment in the existing infrastructure

Maintaining an existing infrastructure requires good information on the performance of systems, whether they are above ground, as with water treatment plant, pumping stations and treated water storage facilities (easier), or underground, such as pipes and valves (more difficult). The difficulty is that systems don't always give early warning of impending failure, and at the point of failure, the systems, especially those underground, are then in an advanced state of decay. Because of this, it has been easy for those responsible for releasing monies for investment to say, 'replacing those items can wait for another year', but that becomes another year and another and another. Equally, it is not an economic proposition to have a policy of replacing all pipes say every 25 years, because age is only one factor, as there is high variability in the life of pipes depending upon a number of factors, including soil conditions and the type of water being distributed. However, it is clearly obvious that an average replacement/refurbishment rate of several hundred years is inadequate. Asset management planning has become one of the most important subjects in water management, beginning with the pioneering work of the UK Water Research Centre (WRc) starting in the late 1970s/early 1980s. Manuals for both sewerage system and water distribution system rehabilitation were produced in 1984 and 1989, respectively. The 4th edition of the Sewerage Rehabilitation Manual, which is still consulted today, was published in 2001, but is now unavailable. Two valuable references are *Legacy: A Decision Maker's Guide to Infrastructure 2025* (Wallsgrove and Cripps, 2025) by Ruth Wallsgrove and Lou Cripps, and *Organisational Change Management for Asset Management: Understanding, Applying and Embedding Asset Management* by Kerr (2025).

So, the most appropriate way ahead is to have good predictive tools for system deterioration and for replacement/refurbishment to take place ahead of the point where the service quality is affected and ahead of the point where action becomes unnecessarily expensive. The costs of doing this are therefore those predicted to maintain the required quality of service on a continuous basis. During the period leading up to privatisation in England and Wales and the production of the first asset management plans, this was referred to as renewals accounting. Unfortunately, standard depreciation accounting systems, used almost universally, are not effective for assets with lives of more than about 10 years, and assume that the value of those assets will diminish over time. Whereas this is true of say a vehicle or a computer, the value of a pipe is the hole it creates in the ground. The value of that hole does not diminish with time, and the important consideration is the cost of maintaining that hole, or rather maintaining its capacity to transmit water without an adverse change in water quality. Therefore, standard accounting systems do not make adequate provision for the replacement of those assets.

Unfortunately, in England and Wales, infrastructure refurbishment was neglected by Governments not putting in place customer service achievement standards and having instead policies unrelated to future water services. The effect now in 2026 is at least 20 years' deficit in infrastructure funding. See chapter 11, England and Wales Story.

Making provision for the poor

The poor in the developing world pay around 15% of their disposable income in coping compared with 1–2% for a piped water supply (Amit and Sasidharan, 2019), The poor do afford water, they have to for survival, but what they would like is a reliable safe supply at a predictable price. What the poor cannot do is save money to pay for an access charge or to pay monthly bills. They need a means of paying an access charge in easy instalments and of paying for water 'as they go'. Most commonly, there are general subsidies which give most benefit to the 'rich', but which are generally insufficient for sustainability, with the result that the service declines and the necessary extensions to distribution systems to serve the poor are not funded. An excellent description of the financial benefit for the poor in having a piped water supply, is given in a diagram by McIntosh (2003) in which he shows how even with a necessary increase in tariffs, the poor are better off than they were when having to pay water vendors. I have slightly modified the diagram in Figure 2.2 to show that, even with the poor increasing their water consumption, which is inevitable when water is more easily available, they are still better off financially. My version of the diagram, which intends to show the principles, does not give costs or quantities, but for numbers in an Asian context, see McIntosh (2003).

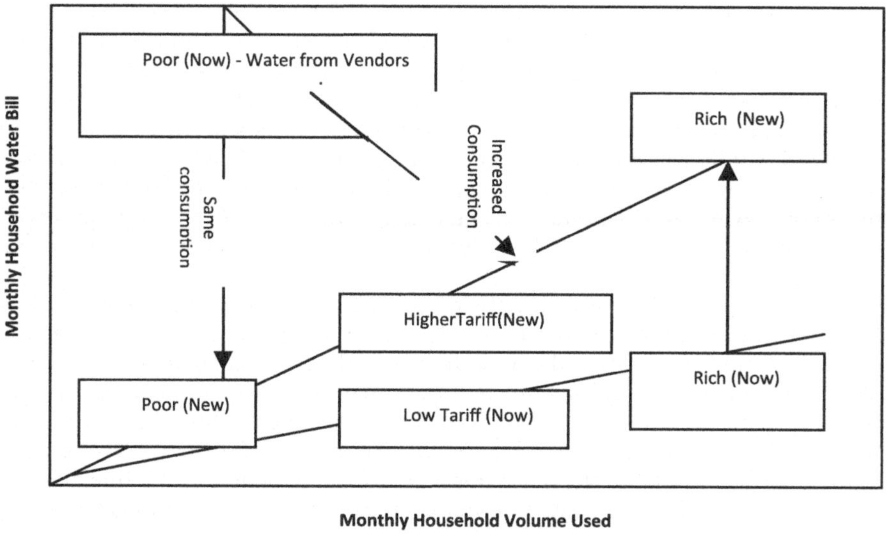

Figure 2.2 Increase tariffs to help the poor. (Source: adapted from McIntosh, 2003).

Low tariffs do not help the poor; on the contrary they deny them a decent water supply. So, what is necessary to help the poor? First, general subsidies should be phased out, with an associated increase in tariffs towards sustainable cost recovery, and available external funds should be directed to refurbishing and extending distributions systems. The refurbishment, coupled with effort to reduce all aspects of non-revenue water, will allow for continuous 24 hour a day supplies, and make water available for those yet to be connected. But refurbishment is expensive and takes time. It would be good if both the donors and the international banks would see such investment as a priority, because additional resources and additional treatment capacity cannot be utilised unless there is adequate distribution capacity.

Second, because the poor are unable to save money, there should be no connection charge, or it should be payable in easy instalments. On installation, consumers should be given advice on how much tap water is used for various purposes, and the cost. But as shown in Figure 2.2, even with an increase in water usage, the bills the poor pay are likely to be lower than when they buy water at a high unit cost from water vendors.

Third, the poor need assistance in how they pay their water bills. One approach is a 'pay as you go' option. This is equivalent to the situation of paying for water from water vendors. One way of paying as you go is the use pre-payment meters. These can be used for payment to kiosk supplies in which consumers buy tokens to provide for access to unmanned facilities, or for use in piped supplies so that consumers can pre-purchase water when affordable. See The Open University (2016). Valuable research on household impacts of water management devices, including water consumption and water pollution, is given in Pereira (2009).

An approach in 2012 (Foster *et al.*, 2012), which was being used extensively in Kenya and Uganda is mobile banking. Mobile phone operators were providing a utility bill payment service through mobile bank accounts. Users could buy credit for their accounts through local agents. Since that time paying by mobile phone has become widespread. As the transaction costs are low utilities can provide for bills to be paid in frequent small amounts. The water consumers benefit as they avoid the inconvenience and cost of travelling to pay through banks or at water utility premises. The water utilities benefit through improved collection and improved cash flow. This leads to an improved water service and greater sustainability.

Another approach to assisting the poor is the use of block tariffs, with a low lifeline tariff for the first quantity of water used. Block tariffs are better than a general subsidy for assisting the poor; however, the financial benefit is gained by all the consumers, including the wealthier citizens. Achieving sustainable cost recovery with a lifeline tariff relies on those consumers who take more water being willing to pay more than its cost, to compensate for the under-recovery at lifeline tariff level. South Africa has a policy of introducing a lifeline amount of 6 kls per month at no cost. In Johannesburg a study (City of Johannesburg, n.d.) concluded that this achieved improved overall revenue collection, as the poorer consumers perceived it to be a fair system resulting in willingness to pay for the amount consumed above 6 kls a month. Israel (Fernandes, 2011) has a block tariff structure providing for 7 kls a month at the lowest level with roughly double the charge for each additional kl. For each family member over four people, an additional 3 kls is available at the lowest tariff. This requires good up-to-date information on household size.

The difficulty with block tariff systems is variability of household size; extended families, which are often in the low-income category, are penalised rather than assisted. Successful systems require something like the approach taken in Israel, with an additional lifeline amount based on household size, but there is a significant administrative burden in establishing and maintaining household records. Without good information on households, larger poor families are probably better off with a flat tariff system, providing that the bills can be paid in easy instalments.

The controversial issue of dealing with non-payment

Cutting off supplies, after due warnings, is commonly applied across the world. For example, in South Africa, government has taken the view that, without this threat, people will not pay their water bills, even with the benefit of a zero cost lifeline system. We know, as mentioned above, that the poor do find the money for water and pay a higher price than those connected to subsidised networks. The priority is how best to provide the poor with easy payment methods. The next requirement is how to implement a cut-off policy so that people do retain access to a safe water supply. In Durban, South Africa (EThekwini Municipality) the first stage following non-payment is to fit a flow restrictor. If the outstanding bill is paid the restrictor is removed. If the bill is not paid the supply is cut-off but a street standpipe supply is made available to provide the free allocation of 6 kls per month.

In Kolhapur (Water and Sanitation Program, 1999) in India, to achieve financial stability, the Mandal, the local community management, introduced a policy of disconnection to ensure that householders pay their bills. This was possible in this system, without accusations of penalizing the poor, due to the availability of public standposts. In the 18-year period up to 1999, only 16 disconnections had been necessary. In comparison, the other district- (i.e. non-community) managed systems in the same area of India have no such policy with a much higher level of non-payment and the risk of unsustainable systems.

One of the benefits of community-managed systems is the community pressure that is applied to achieve universal payment. Probably, the success in Kolhapur was partly as a result of community pressure, and not wholly associated with the threat of cut off.

Responsibility of determining tariffs

A number of the important policy issues have been discussed above and decisions on these are a matter for national or local governments. But having established policies, allowable cost recovery from charges should be set by an independent body which may be called a regulator, which is able to make a wholly objective assessment of operating costs and of the amount of infrastructure refurbishment required for long-term sustainability of the service. The independence of such a regulator can ensure transparency of the charge setting process, through effective consumer participation and through making key documents in the charge setting process available to the public. In this way, political interference in the implementation of policies designed for sustainability, is eliminated or reduced. It is not possible to have an objective method of establishing tariffs to achieve sustainable cost recovery without an integrated economic and water service objectives planning process.

PLANNING

The need for an integrated planning process

A number of conditions are required to achieve sustainable water services. For water supply, there have to be reliable water resources based on defined extreme situations, the necessary water treatment processes to achieve the defined drinking water quality standards, and distribution systems that deliver water in the required quantities to the consumers without adversely affecting its quality. On sanitation, there should be an effective collection system and treatment system, whether or not there is a waterborne sewerage system or a community based recovery system, to achieve the specified environmental objectives. Increasingly, there are linkages between water supply and sanitation through integrated local systems. Whatever approach is taken, there should be clarity of policies, well-defined and realistic drinking water and environmental standards, and a planning system which allows all aspects to be considered in an integrated way.

In addition to new developments, there has to be consideration of the maintenance and refurbishment of existing systems. So often when finance is limited, as indeed it generally is, there has been a tendency to delay investment on the old system, and to make do and mend. Ultimately, continuing deterioration of systems results in loss of serviceability, and failure to meet standards. The requirement is for operational measures that predict when there will be an impact on serviceability so that action can be taken in time. These actions become part of the operational plans.

All of the above aspects have to be financed and the costs recovered. This is why decisions on standards and system development should not be taken in isolation of financing requirements, or in isolation of the ability of the costs to be recovered through grants, subsidies and charges. There needs to be an integrated planning and funding process.

Some examples showing the need for integrated planning

The World Bank North China Study (World Bank (WB) and Ministry of Construction (MOC), China, 2005) included consideration of the institutional governance and regulation aspects of water services in China. A sample of five cities was taken, to give a range of institutional situations and levels of progress in implementing national governments' required reforms as part of the so-called marketization process. Two of these cities, Jinan in Shandong Province and Taiyuan in neighbouring Shanxi Province, had severe water finance problems arising from a lack of integrated planning

In Jinan, in order to protect the city's title of 'City of Springs', the Provincial Government took the decision to stop groundwater abstraction and to build two reservoirs fed by the Yellow River. This was a laudable aim but it was done without effective discussions with the city of Jinan which had to bear the cost of the reservoir developments. The reservoir capacity far exceeded the water resource requirements of the city, thus incurring much higher capital cost than was necessary, leaving Jinan Water Supply Company with a high level of debt.

There was a similar situation in Taiyuan with the Shanxi Provincial Government's decision to embark on a long distance transfer project from the Yellow River. This was necessary due to the need to resolve the problems of over-abstraction of groundwater, but the transfer level greatly exceeded the supply needs of the city. Also, Taiyuan Waterworks Company has been unable to implement the closure of the large abstraction wells providing the major groundwater consumers, as they are regulated at national or provincial level. Both planning and implementation, together with funding and cost implications needed to have been handled in an integrated way. In my comments I include some quote from the report, The World Bank recommended that 'governments at all levels should ensure that the reform of the structure of the water and wastewater sector is carefully designed and implementation is well planned and, in particular develop coherent frameworks for government functions of policy, planning, regulation, asset ownership and business ownership'. This must include the separation of policy, ownership, operations and regulation. The recommendation included clarity on the responsibilities of different agencies and develop agreements and inter-agency procedures for cooperation, particularly for integrated planning across the urban water cycle. 'Establish coherent and clear lines of responsibility and accountability.'

It is common in many countries to rely on effective liaison between various government levels rather than to establish an integrated planning and regulation process. For example, in the United States, standards are set by the Federal Government, environmental standards are regulated at state level, but responsibility for meeting the standards rests with the cities. As a consequence many cities do not see it as their responsibility to be self-funding, and seek federal funds for implementing new regulations and for replacing ageing infrastructure. This attitude has been created partly by federal funding of the introduction of secondary wastewater treatment in earlier years, creating a subsidy mentality. Changing funding policy would require a substantial increase in water service charges, which would be unpopular, even though generally charges are currently very low and the required increases would not result in high household costs in relation to average income.

Governments tend to set drinking water quality standards without regard to how they are going to be achieved. In many countries in Africa, the current quality of drinking water is well below the standards which have been set. The standards are usually based on the WHO guideline values, which is sensible, and countries should aim to achieve those values as a critical public health measure. However, there should be recognition that compliance will take time and require careful planning and investment. There is no point in having standards unless there is effective monitoring, and no point in having monitoring unless there is an enforcement process and the means to plan and fund the necessary improvements. In such cases, there should be recognition of the need for interim standards within a long-term planning process, so that those involved can see a clear way ahead and can measure and celebrate progress as standards are met.

Integrated water and climate adaptation policy including provisions for extreme events

Urban and rural planning should be integrated taking into account rapid urbanisation. Water and sanitation planning should take into account the development of city infrastructure including stormwater management. Importantly, due to uncertainty, particularly related to rainfall, special attention should be given to storage capacity. Rainfall prediction, covering say five-years minimum, is critical and needs to include volume, patterns and severity and be based on probabilistic forecasting. Overall requirement needs to include water supply demand, irrigation requirements and industrial demand.

Catchments should be managed to prevent rapid run-off. Trees consolidate soil, reduce erosion and increase infiltration, and wetlands provide for natural water storage. Buildings and roads should be designed to include rainfall capture and avoid overloading of piped drains and sewer systems. Soakaway swales are not only part of flood management but also increase aquifer storage.

Strategic planning and delivery plans

Typically, a strategic plan would look at a 20-year scenario updated every five years. The plan should be based on a good appreciation of the current situation and based on a realistic view of what is achievable. Often, they are aspirational and serve to show current administrations in a good light. They have to be based on good forecasts of population growth, extent of urbanisation, likely climate change scenarios and uncertainties. Critically, they should refer to international goals but be based on what is sensibly achievable. They should recognise unpopular funding requirements and the need for genuine public consultation, which requires time and effort. A good example is that of Scottish Water (n.d.) whose plans involve extensive public consultation.

Service delivery plans

Service delivery plans are typically for five years, with often a tentative second five-year period. These are detailed costed plans which define what will be done. They cover ongoing operational activities and costs and identify cost projects necessary to achieve the first period of the strategic plan. They are based on the best information available, calculate total investment requirements, and provide the basis for annual budgeting. An Excel spreadsheet can be used to add up all the costs for each year of the plan. The activities of water service providers are normally funded through a combination of tariffs and borrowing.

The service delivery aspects are just part of the overall funding requirement. This is covered in chapter 10, which discusses those aspects which have to be funded from public money, such as water catchment developments, and those where it is feasible to attract private investment, perhaps taking a blended finance approach.

TRANSITIONAL ARRANGEMENTS

The planning requirements covered above provide the necessary framework. The Strategic Plan should be based on a broad appreciation of steps necessary to meet the water bankruptcy challenge including investments required on rainfall catchments. The water service delivery plans relate to the water service providers and should include the transition to increase investment, funding arrangements and staff training.

REFERENCES

Amit R. K. and Sasidharan S. (2019). Measuring affordability of access to clean water: a coping cost approach. *Resources, Conservation and Recycling*, **141**, 410–417, https://doi.org/10.1016/j.resconrec.2018.11.003

City of Johannesburg. (n.d.). Understanding Water Tariffs. https://joberg.org.za

EThekwini Municipality, Water and Sanitation Services. https://www.durban.gov.za/page/water-and-sanitation-services

Fernandes G. (2011). Economic Aspects in Water Management in Israel – Policy and Prices. State of Israel Water Authority, www.water.gov.il/Hebrew/ProfessionalInfoAndData/2012/06-Israel-Water-Sector-Economics-(Policy-and-Prices).pdf

Foster T., Hope R., Thomas M., Cohen I., Krolikowski A. and Nyaga C. (2012). Impacts and implications of mobile water payments in East Africa. *Water International*, **37**(7), 788–804, https://doi.org/10.1080/02508060.2012.738409

Kerr M. (2025). Organizational Change Management for Asset Management: Understanding, Applying and Embedding Asset Management. independently published, ISBN 9798230707677.

Kontula E. (2004). Presentation at World Bank/WaterAid Workshop, August 2004.

Madani K. (2026). Global Water Bankruptcy: Living Beyond Our Hydrological Means in the Post-Crisis Era. UNU-INWEH Report, United Nations University Institute for Water, Environment and Health (UNU-INWEH), Richmond Hill, Ontario, Canada. https://doi.org/10.53328/INR26KAM001

McIntosh A. (2003). Asian Water Supplies: Reaching the Urban Poor. Asian Development Bank and International Water Association, London, ISBN 1 84339 043 4.

Ministry for Foreign Affairs of Finland. (2012). Water Management Lessons in Vietnam. https://um.fi/current-affairs/-/asset_publisher/gc654PySnjTX/content/water-management-lessons-in-vietnam

Pengurusan Aset Air Berhad (PAAB). Annual Reports are available at https://www.paab.my/annual-reports/

Pereira T. (2009). Household Impacts of Water Management Devices – Summary of Initial Research. Presented at EMG Water and climate change seminar, 26 November 2009.

Scottish Water. (n.d.). WaterBriefing. https://waterbriefing.org/home/company-news/item/23745-scottish-water-publishes-ambitious-new-25-year-long-term-strategy

Sustainable cost recovery. International Water Association Reference Paper. https://www.iwa-network.org
The Open University. (2016). Urban Water Supply: Pre-paid meters. https://www.open.edu/openlearncreate/mod/oucontent/view.php?id=80356&section=6.2
UN DESA. (2017). Predicted World Population 2050 and beyond. https://www.un.org/en/desa/world-population-projected-reach-98-billion-2050-and-112-billion-2100
USAID Water and Development Technical Brief 9. (2021). https://bdd.pseau.org/outils/ouvrages/usaid_water_and_development_technical_series_financing_water_and_sanitation_services_2021.pdf
Wallsgrove R. and Cripps L. (2025). Legacy: A Decision Maker's Guide to Infrastructure, independently published, ISBN 9798316742752.
Water and Sanitation Program. (1999). Sustainable Community Management of a Multi-Village Water Supply Scheme in Kolhapur, Maharashtra, India. Field Note SPI Series 2 DFID, London. https://www.ircwash.org/sites/default/files/WSP-1999-Sustainable.pdf
World Bank (WB) and Ministry of Construction (MOC), China. (2005). North China Water Quality Management Study Program. Component E: Regulatory and Institutional Study. https://documents1.worldbank.org/curated/en/281921468019252998/pdf/453250ENGLISH01601PUBLIC10ncwqms1en.pdf
World Population. (2026). https://www.statisticstimes.com>demographics>world-population.php

doi: 10.2166/9781789065527_0023

Chapter 3

Water for communities

Controlling and Reducing Demand

This chapter covers all the aspects of water demand and how it can be reduced. The emphasis is on piped distribution systems, on what the public can do in a disciplined approach to use and conserve water, on governmental controls, and with the vital importance of operation and management of water distribution systems. The content is presented in the following headings:

Part 1: Legislation and Education
- Use water wisely programmes
- Product regulations – maximum limit on toilet flushes

Part 2: Charges – tariff systems
Part 3: Distribution system leakage management
Part 4: Renewable resources

PART 1: LEGISLATION AND EDUCATION

Use water wisely programmes

Water users, under non-drought conditions, tend not to think about their water use behaviour. For example, they wash their hands and clean their teeth under running taps. Cleaning teeth twice can amount to around 20–40 litres a day. Those people without piped supplies would not consider having to collect and carry that amount.

From time-to-time countries or water service providers publicise care in water use. Children are very receptive to such water messages and encourage their families to respond. There are a range from simple posters (see How to Save Water leaflet below) (How to Save Water leaflet) to much broader advice such as use of dishwashers and washing machines. These relate to investing in water saving plumbing such as more-efficient shower heads and spray, and use of grey water (non-toxic) to flush toilets. There is a lot of scope in new building design if people are willing to use waterless toilets and rainwater harvesting (Figure 3.1).

Extensive advice is more generally available for water and energy combined. At the basic end, don't fill a kettle to make one cup of tea. Figures 3.2 and 3.3 are two examples of advice documents (Allen, 2015; Berry, 2008).

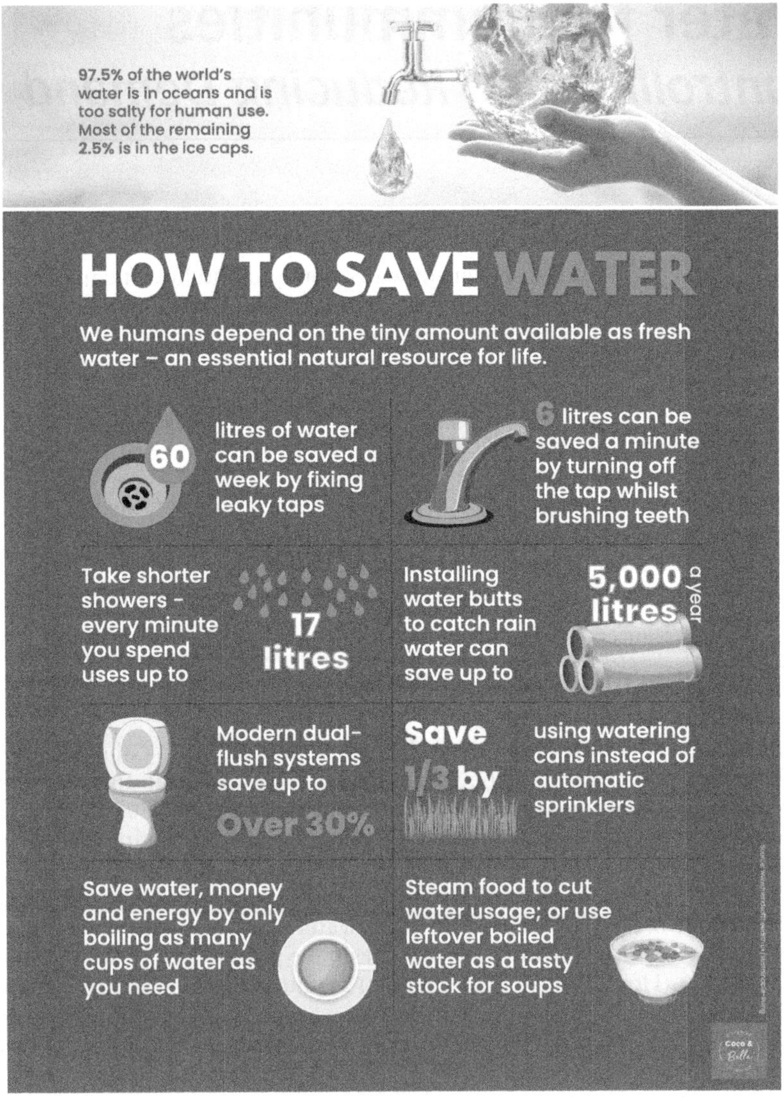

Figure 3.1 Examples of water saving at home. Source: Coco & Bella Creative Design Co., www.cocoandbellacreativedesign.co.uk, used with kind permission.

Figure 3.2 Siân Berry, 50 ways to save water & energy, Kyle Cathie (2008).

Figure 3.3 Laura Allen, water-wise home: How to conserve, capture, and reuse water in your home and landscape, Storey Publishing (2015).

Product regulations

In the UK toilet flushing quantities are defined by the Water Supply (Water Fittings) Regulations 1999 (UK Government, 1999). The maximum toilet flush volume for a water closet toilet pan is 6 litres. Systems in buildings with urinals are limited to a flush of no more than 7.5 litres per hour.

The current United States federal standard is 1.6 US gallons, which is about 6 litres. However, design improvements which meet the WaterSense certification allow a reduction to 1.28 gallons, which is less than 5 litres, thus allowing innovation to save water. Toilet system performance is critical to avoid the use of two or more flushes.

Labelling appliances on water use is becoming more common alongside labels on energy use.

PART 2: CHARGES – TARIFF SYSTEMS

The challenge is to establish charges which are affordable but include an incentive for saving water. Consumer metering is necessary for incentive schemes. The introduction of consumer metering, compared with charges based on say property size, provides for significant benefits.

A method which attempts to reflect types of usage is block tariffs, in which the first block, usually the quantity necessary for basic needs, is low cost, but with increasing tariffs for greater volumes. It is necessary to know the population served from each meter, so not feasible for wider family groups. A useful reference is a book produced by the American Water Works Association (Zieburtz, 2012) (American Water Works Association, 2016).

For financing, timetables and priorities, see chapter 10.

PART 3: DISTRIBUTION SYSTEM LEAKAGE MANAGEMENT

Generally, leakage is the most significant item but frequently given little attention. It becomes the most important 'new' or recovered source of water. A simple calculation demonstrates its importance.

Suppose half (50%) of the water delivered to a distribution system is lost in leakage. Reducing that figure to 30% would mean that 70% would now be available for consumers, an increase of 40%. The alternative of providing an additional source is discussed later in the chapter.

For service delivery viability it is important to consider the wider issue of non-revenue water which includes leakage plus a number of other losses. The starting point is to consider the water balance (IWA Best Practice Water Balance, 2000) (see below). This can be considered akin to a shopkeeper audit. How much is bought in, how much is sold, how much is stolen, how much is damaged and what are the accounting errors. This is critical to the survival of a shopkeeper's business.

System Input Volume	Authorised Consumption	Billed Authorised Consumption	Billed Metered Consumption (including water exported)	Revenue Water
			Billed Unmetered Consumption	
		Unbilled Authorised Consumption	Unbilled Metered Consumption	NonRevenue Water (NRW)
			Unbilled Unmetered Consumption	
	Water Losses	Apparent* Losses	Unauthorised Consumption	
			Metering Inaccuracies	
		Real* Losses	Leakage on Transmission and/or Distribution Mains	
			Leakage and Overflows at Utility's Storage Tanks	
			Leakage on Service Connections up to the measurement point	

Determining the right-hand column is vital. Minimising non-revenue water is a major management challenge.

Commenting on each box:

- System input volume – this can be expected to be measured on the outlet of the treatment works.
- Authorised consumption is in two parts, billed and unbilled.
  - Billed authorised consumption is the amount for which customers receive a bill and is both metered and unmetered consumption. Non-payment can be a significant issue. This can be linked to customer satisfaction, based on issues of good or poor service. Chasing debt is important.
  - Unbilled authorised metered consumption often relates to government offices in which, if supply costs are subsidized, governments choose not to pay for it. This policy is flawed as there is no incentive for governments themselves to use water wisely.
  - Unbilled unmetered authorised consumption relates largely to water for firefighting and any other emergency uses. This is variable year by year and is a judgement based on knowledge of flow from water hydrants and experience.
- Water losses are real (physical) and apparent.
  - Apparent losses include unauthorised consumption. This relates to water that is stolen through illegal connections. There are errors in meter measurements, particularly under-reading at lower flows.
  - Real losses are the total leakage from each part of the distribution system, from large and local distribution mains and property connections.

Measurement of leakage

By making assumptions, an estimate of leakage can be obtained. The water balance is valuable but there can be significant uncertainties about estimating the individual components. Leakage has to be measured. In highly developed supply systems this can be done instantly and continuously by being able to compare metered inputs and all metered outputs. But for most systems the approach is to do night-time input measurements when consumption is at a minimum. Any industrial or commercial users' metered consumption can be subtracted. Measurements can be done zone by zone to identify those areas most in need of leak detection and repair action.

Both overall non-revenue water (NRW) and leakage are important (Farley & Liemberger, 1980; Liemberger & Farley, 1980). Follow-up is dependent upon the initial water balance results. The water balance calculation process, together with the practical measurement of leakage, provide a good indication of the scope for reducing water losses. In situations where physical losses (i.e. leakage) are high, that in itself provides the basis to take action. In situations where leakage is very high, it is not possible to maintain water pressure in the systems and supplies become intermittent. Apart from the poor quality of service, it is unsustainable. See section on intermittent water supplies later in this chapter.

In situations in which NRW is at a level perhaps similar to comparable regions or countries it can be used to determine the priorities for initial action. An important consideration is the quality of information on the systems. Good information on the network and its condition is essential both for the water balance and for leakage measurement. If information on the pipe network is poor or out-of-date, the first investment has to be for contractors using pipe detectors to up-date the maps and where necessary to dig down to determine the size of pipes, their material and condition.

In a Kenyan County study (part of the REACH programme led by the University of Oxford) investment in obtaining system information was proposed in year one of a Service Delivery Plan (see chapter 2 on planning). Pilot applications highlight the lack of basic system information which is vital to allow for improved performance and service delivery.

Case studies on application of the IWA water balance

The approach has been applied widely.

The Asian Development Bank piloted the approach (Asian Development Bank) funding water service providers in five countries, namely Fiji, Georgia, Nepal, People's Republic of China (PRC), and the Philippines. The activities were tailored to each water operator's needs and levels of understanding. For example, in Fiji, the implementation of the AquaRating tool helped the water service provider move from traditional reporting to water balance reporting. In Nepal, the activity was focused on continuing the positive impact of the Water Operators Partnership initiative that led to reducing an NRW of 46–36%.

Yilmaz *et al.* (2021) refer to the applications in Türkiye. The Abstract and Key Findings are reproduced below:

Water losses occurring in distribution systems have an effect on the operating cost, water and energy efficiency, service quality, customer satisfaction, maintenance and new resource demand. The standard water balance recommended by International Water Association (IWA) was used for defining, analysing, regularly monitoring the water balance and sub-components, and determining the weakness and strengths of the system. Water utilities in Türkiye are required to fill the water balance table annually in order to analyse and monitor the performance with the regulation published in 2014. However, important problems have been experienced due to the lack of technical, technological, personnel and economic conditions, data representing the field are not used and in many cases the real performance of the system is not revealed. In this study, the importance of IWA water balance in water loss management and monitoring system performance is emphasized, and the benefits and advantages are detailed by analysing the pilot regions. In addition, problems encountered during filling in the water balance table, mistakes made, incomplete or incorrectly filled data and other problems were discussed. It is thought that this study will constitute a reference for technical personnel in utilities for measuring the data, analysing the components and interpreting the results.

KEY TAKEAWAYS IN TÜRKIYE

- The IWA standard water balance is crucial for analysing and monitoring water loss in distribution systems.
- Türkiye mandates annual water balance submissions to reduce water losses to 30% by 2023 and 25% by 2028.
- NRW encompasses apparent losses, real losses, and unbilled authorized consumptions.
- Field data inaccuracies significantly hinder the effectiveness of the water balance analysis and reporting.
- Active leakage control strategies have successfully reduced water loss from 65–70% to 45–50% in pilot regions.

DISTRIBUTION SYSTEM PRESSURE MANAGEMENT

Pressures in water distribution systems are created by the pressure supplied by the pumps delivering the water, and by the capacity of the network to distribute the water. Leakage levels are directly associated with the average pressures, so it is important to design and operate the systems to the minimum pressure necessary for delivery to the consumers. Zoning and pressure reducing valves and pressure sustaining valves are used for pressure control. Over time networks need to be 'reinforced' to cope with increasing demand. Failure to do so results in increasing pump pressures and loss of pressure in some parts of the system.

Those parts are then entirely without water or there are intermittent supplies. The unsustainability of intermittent supplies is discussed below. There are a number of available network analysis software systems available with support and training, one is EPANet (US Environmental Protection Agency (USEPA) software).

INTERMITTENT SUPPLIES ARE NOT EFFECTIVE AND A VITAL ASPECT FOR INVESTMENT

Intermittent supplies are unsustainable. The shock waves created by de-pressurising and re-pressurising pipes result in an increase in pipe bursts and leaks. Experience in Cyprus is given in Charalambous (n.d.). The paper describes the adverse effects of intermittent supply on customer service and on the integrity of the distribution network, with an increase in the number of bursts and leakage, as well as the financial repercussions to the utility during the water shortage periods faced by the Water Board in the last 20 years. This 'water hammer' effect can be reduced by the use of valves designed to close slowly or in stages.

The result of intermittent supplies is increasing losses which are not sustainable, as generally the deterioration is faster than leaks can be repaired in a whole system, it is necessary to create zones which can be repaired to achieve 24/7 one at a time. A successful example is the modelling of the Amman system in Jordan (El Achi & Rouse, 2020).

The question often raised is surely added additional water supply inputs would solve the problem. This can produce some improvements but without achieving supplies to consumers 24/7. A consultant, Roland Liemberger (now sadly deceased) studied the future of water supply in Dar es Salaam taking a period of 10 years, one in which more and more water supplies were added and the other with extensive refurbishment of the distribution.

The results (see table below) were received by a personal communication – the study has not been published. However, a report by the BBC 24 December 2025 describes 'Chronic water shortages dampen holiday mood in Tanzania's biggest city' – it can be assumed that neither of the two 'solutions' had been implemented and this is the consequence of no action. In practice, it would be necessary to estimate future population growth, additional water required and invest to reduce leakage. Some case studies described later in the chapter have achieved leakage levels as low as 5%.

Leaks and leak detection

Large leaks from say a burst pipe appear on the surface and are easy to locate although expensive to repair. However, most leaks are very small and occur at pipe joints and connections. An interesting statistic illustrates the complexity of the leakage detection. If all the water mains in the UK were placed end to end they would go round the world six times, so imagine the number of

	Unit	Present Situation	After 10 years with Increased Water Input	After 10 years with Concerted Leakage Reduction
System input volume	m^3/day	116,400	258,000	182,000
Increase in input volume	%	–	122	56
Billed consumption	m^3/day	56,951	77,000	145,000
Increase in billed consumption	%	–	35	155
Losses	m^3/day	58,856	181,000	37,000
NRW	%	51	70	20
Average supply time	h/day	10	16	24

Note: The benefit of continuous supply from leak management would be 2½ times more delivered to consumers from only 56% additional water resource.

joints and connections with the possibility of leaking. The propensity to leak depends upon the quality of construction and pipe materials which vary in level of corrosivity and brittleness. Cast iron pipes were used extensively, initially in grey iron which is very brittle, and later as spheroidal graphite (SG) which is ductile. Grey iron pipes laid in a clay soil, which 'heaves' at low temperatures, are most vulnerable to bursts. Replacing grey iron should be a priority for sustainability. Highly publicised loss of supplies, such as the recent ones in South East England (January 2026) and being investigated by the Regulators (*Observer newspaper*, 2026), are likely to have arisen partly from the continuing existence of grey iron pipes. For local water mains, polyethylene pipes are now most commonly used. Steel is used for larger diameter transmission mains, and their durability is dependent on the quality of the coatings in preventing corrosion.

Traditionally, and still in use today are 'listening' sticks. Leak noise correlation equipment is in common use today. The basics, which change very little, are covered in a report Leakage Control Policy and Practice (1980). A review of proactive leakage management technologies used internationally was published in 2007: Leakage Management Technologies (Leakage Management Technologies, 2007).

A leak makes a distinctive sound pattern which can be detected in any direction along a water main. Such sound patterns can be picked up by microphones and measuring the time difference between the signals at two points. By signal correlation the exact location (within a metre) can be determined. See figure below.

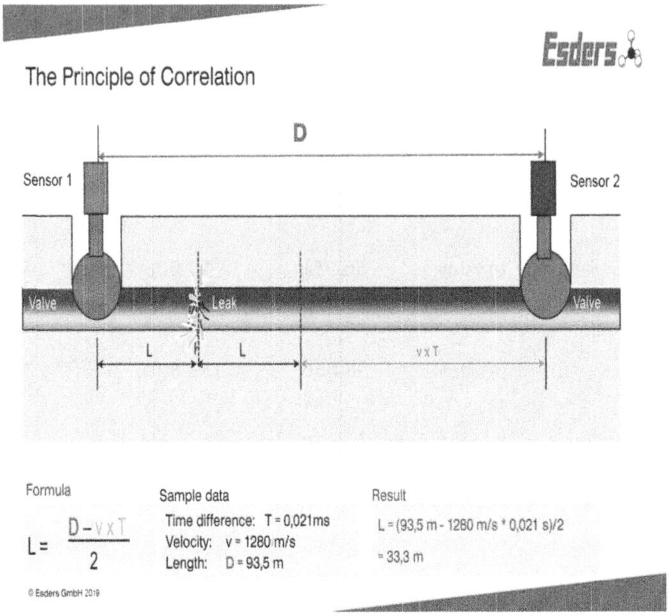

Modern correlation uses radio pick up and fits in a backpack. It is an essential equipment to achieve efficient leak location and repair.

The most advanced distribution systems, which are managed in real time, include continuous sound correlation monitoring such that leaks are picked up and located as they occur (see Singapore case study).

LEAK MANAGEMENT CONTRACTS

Leak detection and repairs are often subcontracted out to specialists. It is one aspect in which there can be performance-related contracts. Contractors are paid by measured leak reduction results (Kingdom *et al.*, 2018).

WATER LOSS CASE STUDIES

Five case studies are discussed.

(1) Phnom Penh, Cambodia
   - Leakage reduction as part of water service delivery reform
   - (leadership by an exceptional person)
(2) National Water and Sewerage Corporation (NWSC) Uganda
   - Another water service delivery reform example through a business-management approach
(3) Singapore
   - A comprehensive approach to NRW – down to 5%.
(4) Copenhagen, Denmark
(5) The Netherlands

PHNOM PENH, CAMBODIA

After Pol Pot and the Khmer Rouge were driven out of Cambodia, there was a limited water supply service and high losses. Ek Sonn Chan, who was an engineering graduate, managed to survive the Khmer Rouge killings by pretending to be a farmer. On independence he was given freedom by the new government to develop the water service and received financial support from the French Government. In 13 years he had achieved a world class water supply service operation. A summary of the results is given in the following table.

Performance Indicator	1993	2006
Coverage area (%)	25	90
Number of connections ('000 s)	27	147
Supply duration (h/day)	10	24
Revenue collection (%)	48	99.9
NRW (%)	72	8
Staff per 100,000 connections	22	4
Financial situation	Subsidy	Full cost recovery

Ek Sonn Chan was then appointed as Under-Secretary of State for Water at the Cambodian Ministry of Industry, Mines and Energy to apply his experience across Cambodia. In 2024 he was appointed Chair of the Phnom Penh Water Authority. This is an example of what an exceptional manager can achieve with political support (Biswas & Tortajada, 2010; Chan, 2009; Leong, 2010).

NWSC UGANDA

This example is the transformation of the national water service following a new board of NWSC and the appointment of a new manager William Mulhairwe. William identified the problems as poor customer service, poor organisational culture, large debts, low cost recovery, high NRW and corruption amongst staff. The approach was to delegate responsibility to the area managers, who were given performance contracts. A benchmark approach was used to stimulate competition between areas with areas rewarded for best improvements. There was extensive training on best practice (see results below) (Mulhairwe, 2009).

Performance Indicator	1998	2011
Service Coverage (% of population served)	48	75
Total piped connections	50,826	272,406
New Connections per year	3,317	25,000
Household metered Connections	37,217	271,734
Staff per 1000 Connections	36	6
Collections Efficiency*	60%	98%
Non-Revenue Water	60%	33%
Metered Accounts	65%	99%
Annual Turnover (USD million)	8	51
Profit* (USD million)	−3 (Loss)	12 (Surplus)

It is interesting to consider whether there was continuing performance in NWSC after William Mulhairwe's departure in 2011. With data taken from the NWSC 2021/2 Annual Report (National Water & Sewerage Corporation Integrated Annual Report 2021/2), the growth in services has been sustained but NRW remains similar to that of 2011, suggesting that there remains scope for securing some of the water being lost (see table below).

Key Performance Indicator	FY 2017/18	FY 2018/19	FY 2019/20	FY 2020/21	FY 2021/22
Water Services					
Number of Towns (Nos.)	236	253	258	258	262
Water Service Coverage (%)	83.7	83.7	84	84	84
Water Production (Million m^3)	126.2	134.9	140	144	164
Water Services					
Water supply (Million m^3)	120.1	127.8	132	137	158
Water Sales (Million m^3)	82.9	90.5	87	90	102
Non-Revenue Water (NRW, %)	31	29	34	35	35
New Water Connections (Nos.)	50,341	69.215	61.521	51.788	54,697
Connections taken over with New Towns	12,865	2,079	3,328	–	26856
Total Water Connections (Nos.)	582.863	659.157	724,006	775.794	857.347
New Public Stand Pipes (Nos.)	3.342	3.550	4.429	3,267	2.141
Total Public Stand Pipes (Nos.)	12,305	17,186	21,600	24,867	28,858
New water mains extensions (km)	2,021	2,727	2.135	439	1,305
Extensions taken over in New Towns (km)	700	430	216	77	0
Total water pipe network (km)	14,466	17,556	19.974	20.495	21.794

SINGAPORE

Reducing NRW was part of the Lee Kuan Yew (President) drive to achieve independence from Malaysia on water resources. The programme covered all aspects, pressure management, leakage measurement, extensive leak detection, a rapid repair programme and minimising meter error, which achieved NRW of just 5%.

It sets up and organises an annual event, Singapore Water Week, bringing together the most developed ideas for development.

An example of its development is the Introduction of sensors and online continuous acoustic leak detection on the large transmission systems for early identification of large leaks. See reference on Singapore enhances transmission main leak detection with next-generation physics-driven analytics (Singapore transmission main leak detection).

Part of Singapore's ongoing development is consideration of how artificial intelligence might be used to improve operational performance.

For details of the initial development, see The Singapore Water Story (Tortajada *et al.*, 2013).

COPENHAGEN, DENMARK

Water matters are decentralised in Denmark with supply services and water resource provision handled by regional and local authorities. To stimulate action on leakage, service providers with more than 10% NRW were subject to penalty charges from 1993. Good progress was made over the period 1975–2020. In 2015 the service providers entered into a commercial agreement. The LEAKman Concept (*The LEAKman Concept*) was developed by an association of nine Danish partners representing technology providers, consultants, water utilities and the Technical University of Denmark. See 1975–2020 leakage data below.

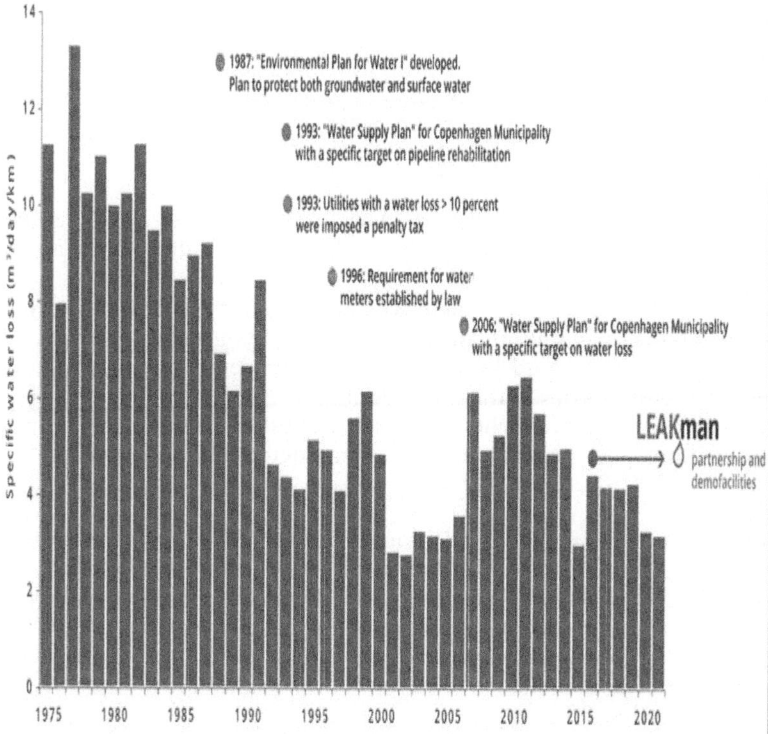

THE NETHERLANDS

Success in managing non-revenue water in the Netherlands stems from an active government with national plans. The government has worked closely with the national water utility association Vewin with consistent policies related to an efficient water supply service. In Amsterdam NRW is 5% (https://en.wikipedia.org/wiki/Water_supply_and_sanitation_in_the_Netherlands).

PART 4: RENEWABLE RESOURCES

Treated wastewater

Advanced treatment of wastewater has been shown to be safe to drink and is increasingly being pursued in dry areas. Some countries have specific quality regulations. EU treated wastewater regulations were introduced in 2023 (EU Treated Wastewater Regulations, 2025). Experience has shown that extensive consumer consultation is necessary to address the 'yuk' factor.

Such consultation was missing in proposals to use treated sewage in England. The tabloid press destroyed the proposal with 'sewage on tap' headlines. This forced the company at the time to discharge the wastewater treated to drinking water standard to the river, which was then abstracted downstream as river quality water for re-treatment. Progress on attitudes has since been made in the UK and treated wastewater is now being seen as an essential water resource.

Current required treatment technologies involving membranes removes minerals which for health reasons need to be replaced.

Situations in which the treated wastewater is one part of the available water, such as in Singapore, the blended water contains sufficient minerals.

Case studies:

Windhoek, Namibia, in southern Africa: (Lahnsteiner *et al.*, 2024). The system has now been operating for over 40 years. Windhoek has an average rainfall of 370 mm per year.

Singapore: A very open and effective approach achieved public acceptance. This involved the formation of an expert advisory group of world experts. Also, in parallel with the treated processes is a visitor centre which explains each process part, with bottles of the final product labelled 'New Water' provided to taste and drink. Groups of school children and other groups are taken through the centre (Singapore transmission main leak detection).

Desalination

Historically, in arid areas countries such as Saudi Arabia, with no freshwater, rely on treated seawater. The distillation process was used which requires a lot of heat energy so its use was subject to there being a sustainable energy source. Increasingly, reverse osmosis (RO) is being used which requires electrical energy with the prospect of the use of renewable energy sources such as solar power. Solar power can also be used for multiple small local units. See general reference (Desalination).

In 2026, there are signs that desalination is being considered in water-stressed cities. The Brihanmumbai Municipal Corporation has entered into a

20-year contract to build, operate and maintain a seawater RO desalination plant to provide Mumbai city with 200 million litres of drinking water every day. (Source: IWA Magazine 'The Source', April 2026.)

Transition arrangements

Reducing demand will become even more important. As part of public communications information should be provided on water saving behaviour. Leaflets such as those presented in this chapter provide a good source of information.

Planning water distribution leakage programmes together with associated training should be a priority as it will take time for the operations to be put in place.

REFERENCES

Allen L. (2015). Water-Wise Home: How to Conserve, Capture, and Reuse Water in Your Home and Landscape. Storey Publishing, New York, USA. ISBN 978-1612121697.

American Water Works Association (2016). Principles of Water Rates, Fees and Charges, 7th edn, American Water Works Association, Denver, Colorado, USA. ISBN 9781625761910.

Asian Development Bank: https://www.adb.org/results/pilot-application-iwa-water-balance-methodology

Berry S. (2008). 50 Ways to Save Water & Energy. Kyle Cathie, London, UK. ISBN 9781856267731.

Biswas A. K. and Tortajada C. (2010). Water supply of Phnom Penh: An example of good governance. *International Journal of Water Resources Development*, **26**(2), 157–172. https://doi.org/10.1080/07900621003768859

Chan E. S. (2009). Bringing safe water to Phnom Penh's city. *International Journal of Water Resources Development*, **25**(4), 597–609. https://doi.org/10.1080/07900620903306323

Charalambous B. (n.d.). The Hidden Costs of Resorting to Intermittent Supplies. https://hydrocontrolltd.com/wp-content/uploads/2025/07/The-Hidden-Costs-of-Resorting-To-Intermittent-Supplies.pdf

Desalination. https://www.britannica.com/technology/desalination

El Achi N. and Rouse M. J. (2020). A hybrid hydraulic model for gradual transition from intermittent to continuous water supply in Amman, Jordan: A theoretical study. *Water Supply*, **20**(1), 118–129. https://doi.org/10.2166/ws.2019.142

EU Treated Wastewater Regulations (2025). https://environment.ec.europe.eu

Farley M. and Liemberger R. (1980). Developing a Non-Revenue Water Reduction Strategy, Part 2: Planning and Implementing the Strategy. Available at: https://sswm.info/sites/default/files/reference_attachments/LIEMBERGER%20FARLEY%20 2004%20Developing%20a%20NRW%20Reduction%20Strategy.pdf

How to Save Water leaflet: digital downloads and printed poster versions are available from Coco & Bella Creative Design Co., www.cocoandbellacreativedesign.co.uk

https://en.wikipedia.org/wiki/Water_supply_and_sanitation_in_the_Netherlands

IWA Best Practice Water Balance (2000). see https://www.leakssuitelibrary.com/iwa-water-balance/

Kingdom B., Sy J. and Soppe G. (2018). The Use of Performance-Based Contracts for Nonrevenue Water Reduction: Output of the Global Program on Developing Good PBC Practices for Managing NRW. World Bank, Washington DC, USA.

Lahnsteiner J., Honer T., Ashipala L., Nikodemus K., Poussade Y. and van Rensburg P. (2024). Windhoek/Goreangab direct potable water reuse, case study Namibia, Handbook of Water and Used Water Purification, 1301–1318, Springer, Cham. https://doi.org/10.1007/978-3-319-78000-9_170

Leakage Control Policy and Practice (1980). Published by Water Authorities Association. London, UK. ISBN 0 904561 95 X, https://www.dwi.gov.uk/research/completed-research/drinking-water-distribution-networks/leakage-control-policy-and-practice/

Leakage Management Technologies (2007). AWWA Research Foundation and IWA Publishing. https://database.waterwise.org.uk/wp-content/uploads/2019/09/EPA-2007_Leakage-Management-Technologies.pdf

Leong C. (2010). Ek Sonn Chan and the transformation of the Phnom Penh Water Supply Authority. *SSRN Electronic Journal*, https://doi.org/10.2139/ssrn.1588071

Liemberger R. and Farley M. (1980). Developing a Non-Revenue Water Reduction Strategy, Part 1: Investigating and Assessing Water Losses.

Mulhairwe W. T. (2009). Making Public Enterprises Work. IWA Publishing, London, UK. ISBN 9781843393245

National Water and Sewerage Corporation Integrated Annual Report 2021/2.

Observer Newspaper. (2026). London, UK.

Singapore transmission main leak detection https://h2oglobalnews.com/singapore

The LEAKman Concept. Denmark, https://leakagemanagement.net/

Tortajada C., Joshi Y. and Biswas A. (2013). The Singapore Water Story: Sustainable Development in an Urban City State. Routledge, London, UK. ISBN 978-0-415-65783-9

UK Government (1999). The Water Supply (Water Fittings) Regulations 1999, available at https://www.legislation.gov.uk/uksi/1999/1148/contents

US Environmental Protection Agency (USEPA) software, EPANET Application for Modeling Drinking Water Distribution Systems, available at https://www.epa.gov/water-research/epanet

Yilmaz S., Özdemir Ö and Firat M. (2021). Application of IWA standard water balance in strategic water loss analysis: benefits and problems. *Environmental Research and Technology*, **4**(2), 176–183. https://doi.org/https://doi.org/10.35208/ert.886829

doi: 10.2166/9781789065527_0039

Chapter 4

Irrigation
Living beyond our means

Irrigation is critical for food production. In volume, it accounts for around 65–70% of world water use. For this reason, it becomes even more important with water bankruptcy as the rainfall deficit must be addressed to reduce worldwide famine.

Historically, water for irrigation was largely rainfall and wells, taken for farming with no charges. Rainfall harvesting, using contour bunding, collected seasonal rainfall for irrigation and groundwater recharge in for example India (Chiturike *et al.*, 2024). Colonials, with large farms, built 'small' reservoirs using dams for irrigation for example South Africa (AgrifoodSA magazine). Climate change, with more frequent and more severe droughts, requires more efficient use, greater storage, use of treated wastewater and desalination.

Farmers are often strong politically and can have adverse influence on conservation measures. For example, the French government reacted to the 2022 European drought by using groundwater to create irrigation ponds for farmers (Renault, 2023).

MANAGING IRRIGATION WATER

There are a number of ways of reducing water demand. These are listed below and discussed in turn:

- Crop selection
- Rainwater harvesting and groundwater recharge
- Controlled abstraction
- Methods of irrigation
- Alternative sources
  - Treated wastewater for irrigation
  - Desalination

CROP SELECTION

Poor selection can be disastrous. One extreme example is unsustainable, that of cotton growing in Uzbekistan and the 'loss' of the Aral Sea (Hanks, 2021).

There is scope for significant water demand reduction by selection of crops, for example changing from one cereal crop to another requiring less water (Chakraborti *et al.*, 2023). A team from Mumbai and other universities studied irrigation on the Indo-Gangetic Plain (IGP), the so-called food bowl of India with a population of 400 million.

Increased cereal production coincided with over-abstracted groundwater The team addressed the following questions:

(i) What are the trends in cereal production and water consumption?
(ii) What cereal crop substitution would maximise calorie production and farmers' profits and minimise water consumption?
(iii) Which combination of irrigation practice (flood and drip irrigation) and crop switching maximises groundwater net recharge?

They used the WaterGap model which computes water flows and storages as well as water withdrawals and consumptive uses on all continents. It showed an increasing trend in water consumption for cereal crops in the IGP, but without any increase in calorie production. There was an overall increase in water consumption and reduced groundwater recharge. The team developed a multi-objective optimisation model with the objectives of reduced water consumption, increased crop yields and farmers profits with the fractions of cereal-grown areas as decision variables. The model found that switching from rice to millets (pearl millet) and sorghum in the Kharif (monsoon) season and from wheat to sorghum in the Rabi (winter) season could potentially reduce water consumption by 32%, improve calorie production by 39% and increase farmers' profits by 140%, and switching crops offers a larger reduction in groundwater depletion and energy savings than improving irrigation efficiency (i.e. from flood to drip irrigation).

RAINWATER HARVESTING AND GROUNDWATER RECHARGE

Natural groundwater recharge historically has provided sustainable water resources, although in developed civilisations, collection and storage systems were built as part of city development. However, over the last 100 years, with rapid population growth and urbanisation, aquifers have been over-abstracted, and control of run-off has been neglected with resulting flooding. The problems being compounded now by more extreme events. There are many aspects to the problems and the solutions. One is the re-introduction of wetlands. Another is planting of trees. Others are controlling the run-off from buildings, and the use of porous paving in cities; and artificial recharge in the catchment with collection ponds, checkdams and roadside swales.

In Sri Lanka Tank Cascade Systems (TCSs) (Sri Lanka Tank Cascade Systems) were extended in the 12th century AD onwards with a visionary approach to water conservation. Later, many of these systems became neglected due to the influence of 'colonials' who introduced 'modern' systems. However, at the end of the 20th century, with population growth, traditional rainwater systems were included in the development programmes. There is now the recognition of the need for 'blended' approaches.

Historically, the Alwar District of Rajasthan, India had effective rainwater harvesting, but the British introduced 'urban-type' infrastructure systems relevant to European cities, but not appropriate for the requirement in India. After India had become independent, the systems deteriorated with over-abstraction. Village elders, assisted by an NGO, Tarun Bharat, decided to re-introduce rainwater harvesting. The story in summary is the construction of low earth walls built along contours which collected rainfall forming shallow reservoirs called johads. In total, 4500 such johads were built across the 1000 villages involved. The collected water gradually permeated through the ground leaving fertile soil and gradually the previously dry springs began to flow, leading to the previously dry Arwari river to begin to flow. It was critical that the recreated sustainable environment was not lost and the democratic structure in India allowed for the villages to form an organisation, the Arwari Parliament, to manage the river (Hasnat, 2005; Jal Bhagirathi Foundation; Sarkar, 2018).

CONTROLLED ABSTRACTION

Over-abstraction of groundwater is a major concern in many parts of the world – the climate change impact on rainfall predictions is exacerbating the problem requiring the need for better science to establish management policy. A study by the Institute for Hydromechanics and Water Resources Management, ETH Zürich, Switzerland (Kinzelbach *et al.*, 2003) addressed this concern. The study concluded that the sustainable management of aquifers is a burning problem in many countries. New scientific tools can assist in its solution, both in the analysis of whether the present management is sustainable and in the definition of strategies and measures to achieve sustainability. Models will always play a role in this task. A new model generation can use new data sources such as environmental tracers, remote sensing data and geophysical data. It must however not only simulate an average deterministic situation but also analyse the uncertainty of its predictions by a stochastic approach in order to remain credible. Finally, the natural and engineering sciences have to interface much more with economics and politics to be of real practical use. Irrigation being the major groundwater consumer requires the 'political' influence of farmers to be addressed.

METHODS OF IRRIGATION

Irrigation methods include surface, sprinkler, drip and subsurface techniques.

Surface irrigation uses gravity for flood/furrow watering, while drip irrigation applies water directly to the roots of plants which is highly water efficient. Sprinklers are the same as rainfall whereas subsurface irrigation minimises evaporation through water going directly below the surface.

ALTERNATIVE SOURCES

Wastewater for irrigation

Historically community wastewater has been widely used for agriculture with growth benefits arising from the nutrients in the wastewater. Two case studies in Nepal included water sampling and interviews with farmers. The main findings in 2006 were that the farmers were finding the practice valuable in the dry season with the wastewater providing greater water security across seasons. However, the quality of water used ranged from wastewater diluted by river water to raw sewage with some farmers experiencing health problems. The authorities had little awareness of the practice and there were no regulations. There are no updates on the experience but the Government plans to have all wastewater treated by 2030 (World Bank, 2014).

Irrigation

This is a rapidly developing scene – see chapter 3 for developments related to drinking water. For irrigation it may be limited to areas with no fresh water and having available energy resources to treat sea water. For irrigation generally it is likely to be too expensive although it is possible that there may be some high value crop applications.

VIRTUAL WATER

Virtual water is the total volume of water used to produce a product or service throughout its entire supply chain, from raw materials to the final product. The example often quoted is that it requires 14 litres of water to produce one cup of coffee. The term virtual water was invented in 1993 by the late *John Anthony Allan*, a British geographer and an emeritus professor at King's College London. In 2011 he expanded on his concept in a book on Virtual Water (Allan, 2011). Although it had some impact at the time with the concept being taken up by the World Bank, the EU, the US government and the WWF, nothing was done to address the resulting water deficit problems in the food producer countries. In todays' era of water bankruptcy it will have to be taken seriously as the producer countries' water deficit is becoming critical.

The problems faced are immense. The economies of the producer countries have become dependent on the sales of products, such as coffee. Would the purchaser countries be willing to go back to the situation when only locally produced products were available – this is highly unlikely. Big businesses such as supermarkets are dependent on meeting consumers' demands – what can they do to stimulate change? To what extent can the consumer countries compensate the producer countries by investing in water saving measures? Are there variants of say coffee beans requiring less water? Can we expect governments to address the problems?

TRANSITION ARRANGEMENTS

Universities and other research organisations should be funded and encouraged to extend studies on crop selection to reduce the water requirement. Water demand labelling should be introduced widely. Water bankruptcy public awareness programmes must be established. The required changes highlighted by the questions related to virtual water will take a number of years and governments should establish a group of all key players including supermarkets to develop a way forward.

REFERENCES

AgrifoodSA magazine. AgrifoodSA magazine, 20 March 2023. Types of Water Reservoirs and Their Uses, https//farm.co.za

Allan T. (2011). Virtual Water: Tackling the Threat to Our Planet's Most Precious Resource. I.B. Tauris, London, UK, ISBN 9781845119843.

Chakraborti R., Davis K. F., DeFries R., Rao N. D., Joseph J. and Ghosh S. (2023). Crop switching for water sustainability in India's food bowl yields co-benefits for food security and farmers' profits. *Nature Water*, **1**, 864–878, https://doi.org/10.1038/s44221-023-00135-z

Chiturike P., Gotosa J., Nyakudya I. W., Madamombe S. M., Mandumbu R., Chirinda N., Kugedera A. T. and Nyamadzawo G. (2024). The effects of contour-based rainwater harvesting and integrated nutrient management on maize yields in semi-arid regions of Zimbabwe. *CABI Agriculture and Bioscience*, **5**, 41, https://doi.org/10.1186/s43170-024-00233-2

Hanks R. R. (2021). Louder than words: a profile of the destruction of the Aral Sea and its consequences. *Education About Asia, Teaching Resource Journal*, **26**, https://doi.org/10.65959/eaa.1685

Hasnat S. M. (2005). Arvari Sansad: the farmers' parliament, ICICMOD, https://lib.icimod.org/records/j8vnw-bj111

Jal Bhagirathi Foundation. Jal Bhagirathi Foundation website, https://www.jalbhagirathi.org

Kinzelbach W., Bauer P., Siegfried T. and Brunner P. (2003). Sustainable groundwater management – problems and scientific tools. *Episodes Journal of International Geoscience*, **26**(4), 279–284, https://doi.org/10.18814/epiiugs/2003/v26i4/002

Renault M. (2023). French people are fighting over giant pools of water: the underground reserves that fill mega-basins are not an infinite resource. The Atlantic, 7 June 2023, https://www.theatlantic.com/science/archive/2023/06/water-mega-basins-reservoirs-france-drought/674313/

Sarkar J. (2018). Community driven water resource management: a case study in Alwar District, Rajasthan, India. *International Journal of Research and Analytical Reviews*, **5**.

Sri Lanka Tank Cascade Systems. Sri Lanka Tank Cascade Systems, https://en.wikipedia.org/wiki/Tank_cascade_system

World Bank. (2014). Nepal: Irrigation and Water Resource Management, https://www.worldbank.org/en/results/2014/04/11/nepal-irrigation-and-water-resource-management

doi: 10.2166/9781789065527_0045

Chapter 5

Water for industry – managing demand and quality

Water is used in many ways in industry, some of which is returned to the environment. Demand management and impacts on water conservation and quality are considered for each of the following:

- Water for power station cooling
- Water for hydropower
- Water for chemical industry
- Water for textile industry
- Water for mining
- Water for the metal industry
- Water for food processing
- Water for construction

WATER FOR POWER STATION COOLING

Once-through systems water from nearby sources (e.g. rivers, lakes, aquifers, or the ocean), circulate it through pipes to absorb heat from the steam in systems called condensers, and discharge the now warmer water to the local source. No water is 'lost' but there can be higher densities of fish just downstream of power stations. However, depending on the screening, fish can be trapped on the water intakes to the plant. The heat from nuclear power station cooling (Hansa Engineering) can be used for town district heating systems. In France and other countries which have extensive nuclear power electricity generation, it provides for substantial savings in use of fossil fuels for industrial and domestic heating. Extracting heat from 'conventional power station' water cooling is less feasible because it is related to the efficiency of the steam turbine generation process.

WATER FOR HYDROPOWER

No water is lost in energy generation from hydropower – it uses its kinetic and potential energy. However, hydropower can limit the amount of water which can be used for general use if it is critical to maintain sufficient flow to drive the turbines. There are many examples of this combined water and energy dilemma which is being worsened by climate change: (Opperman, 2022). Hydropower from the Kariba Dam provides half of electricity needs for Zambia and Zimbabwe. The falling level of water in the reservoir has resulted in significant power cuts. Falling levels in the Mississippi and Colorado rivers in the United States are causing concerns. Due to an extended drought, hydropower which provides 80% of electricity to Sichuan province in China, has been reduced to half.

WATER FOR CHEMICAL INDUSTRY

Water for the chemical industry includes the production of 'commodity chemicals', from pharmaceutical products to polymers, food, petrochemicals and intermediates, and dyes. It is the major water user in industry and often requires highly treated water. Demand is predicted to increase by 400% by 2050, so a candidate for attention in relation to the UN Water Ministers Goals 2023 (UN 2023 Water Conference) of 'Overcoming the dependence on ever-rising water consumption'. In the water bankruptcy era, this has become more urgent. As there are so many different industrial sectors, it is too large a subject to address in this outline consideration of the various elements. As a general point, as it requires extensive water treatment anyway, it will be necessary for major investment in water re-use.

WATER FOR TEXTILE INDUSTRY

This is another major area for both water demand and water pollution. Cotton production water demands are high – to produce a single cotton T-shirt requires 1700 litres of fresh water. The virtual disappearance of the Aral Sea was due to cotton production demand on a very large water lake for which there was little water recharge. The solution is not to use synthetics instead due to their carbon footprint and concerns over plastic microfibres. Currently only 1% of cotton is recycled for new garments, but there is a growing trend in buying second hand clothes. Increased attention is being given to reduce the fashion demand for new clothes, but with implications for those producing countries on income and employment. Given the commercial interests of the garment market, both public education and regulation will be necessary. The European Union is taking this area seriously as part of pursuing a circular economy (European Parliament).

WATER FOR MINING

Mines owned by the five largest mining groups have 70% of their mines in water-stressed countries. The International Council of Mining and Metals (ICMM), together with the International Finance Corporation of the World Bank (IFC), in recognising the growing problem have established a collaborative approach which they label 'Shared water; shared responsibility; shared approach'. ICMM has produced a number of guidance documents (ICMM Guidance Documents).

- ICMM's practical guidance on a catchment approach to water management.
- Water, mining and communities.
- The Value of Water – produced by the World Wildlife Fund (WWF)
- ICMM's guide to consistent water reporting

An example is given of Mongolia's Gobi Desert area where mining is critical to the country's economy. In one mine water consumption has been reduced to half of that used by comparable mines elsewhere. However, the whole area is subject to severe challenges on water sustainability (Purevjav *et al.*, 2025).

WATER FOR METAL INDUSTRY

Water is used for cooling and for various metal processing requirements. Most water is recovered but, as it is highly polluted by some metals such as cadmium and the processing chemicals used, extensive wastewater treatment is required. The effluents are required to meet regulatory standards. See Heavy Metal Emissions in Europe (European Environment Agency, 2025) published by the European Environmental Agency on 13 September 2024.

WATER FOR FOOD PROCESSING

Food production requires a significant volume of water, with around 25,000 litres required to grow and produce a day's supply of food for a family of four. Most water is required to meet WHO drinking water quality guidelines. For hygiene purposes, food and beverage manufacturers tend to use water once before discharge. However, with higher water prices, there is an increase in recycling and this will become more important in the challenge of water bankruptcy. Some water goes into beverages and bottled drinks, but most water is discharged either to the environment. A processing poultry project in Alabama (PhD Project) demonstrates both the need and value of wastewater regulation. A PhD student at the University of Alabama developed a pilot plant which included watering lettuce plants with treated poultry washing effluent. Having that water available in the water bankruptcy era would be highly beneficial.

WATER FOR CONSTRUCTION

On construction sites, it is now being recognised of the need for water saving. In the UK the Chartered Institute of Building (CIOB) advises that 'Construction must take water scarcity seriously' and promotes good practice (May 2024) (Chartered Institute of Construction (CIOB), 2024). Its members are advised to have a water management plan, have someone designated on site as water awareness champion, have a meter measuring site consumption to identify leaks, identify opportunities for use of non-potable water for example vehicle washing and use trigger guns on hoses to ensure that hoses are not left running. In the UK, since 1990, it has been mandatory to have a water meter installed on newly built residential properties. This practice should be used worldwide.

REFLECTIONS ON WATER FOR INDUSTRY

Water for industry is significant in the overall water demand picture. There are many sectors, all with different combinations of demand and quality issues. In the examples given, industry associations have recognised the importance of water scarcity, often driven by the risks to business. Water saving steps can be pursued through education and self-interest, but regulation plays an important role. There is good scope for reducing water demand in dealing with the challenges of water bankruptcy.

TRANSITION ARRANGEMENTS

Governments in their initial assessment of the scope for water demand savings should include the examples given in this chapter. A large number of individual businesses will be involved but industry associations can provide assistance in planning.

REFERENCES

Chartered Institute of Construction (CIOB). (2024). Construction must take water scarcity seriously, https://www.ciob.org

European Environment Agency. (2025). Heavy Metals Emissions in Europe, https://www.eea.europa.eu/en/analysis/indicators/heavy-metal-emissions-in-europe

European Parliament. European Parliament, https://www.europarl.europa.eu/resources/library/images/20230613PHT98336/20230613PHT98336_original.png

Hansa Engineering. How do combined heat power plants (CHP) create warm water for district heating networks? Hansa Engineering, https://hansa-engineering.se/en/knowledge-base/how-do-combined-heat-power-plants-create-warm-water-for-district-heating-networks/

ICMM Guidance Documents. ICMM Guidance Documents, https://www.icmm.com

Opperman. (2022). Hydropower and Water Scarcity: the growing climate risks of a climate solution. Forbes, https://www.forbes.com/sites/jeffopperman/2022/12/16/hydropower-and-water-scarcity-the-growing-climate-risks-of-a-climate-solution/

PhD Project. A personal communication on an unpublished PhD project in Alabama, USA.

Purevjav B., Klein B., Dierkes J., Kunz N., Xavier A. and McFaul S. (2025). Enhancing water sustainability in the Gobi Desert: processes based on IWRM principles. *Discover Water*, **5**, 17, https://doi.org/10.1007/s43832-025-00204-2

UN 2023 Water Conference. UN 2023 Water Conference, https://sdgs.un.org/conferences/water2023

doi: 10.2166/9781789065527_0049

Chapter 6

Water catchment and urban water management

INTRODUCTION

This chapter is in three parts:

Part 1: Catchment management
Part 2: Urban water management
Part 3: Water storage

In simple terms, to manage the impact of water bankruptcy it is necessary to capture more rainwater, store what is necessary to balance out annual rainfall variations, and reduce water consumption in communities, for irrigation and by industry. Chapters 3, 4 and 5 are devoted to demand reduction. Parts 1 and 3 of this chapter deal with catchment management to capture more rainfall and approaches to water storage. Part 2 has some relevance to storage but relates largely to managing the water in communities in relation to flood aspects and the quality of receiving waters.

PART 1: CATCHMENT MANAGEMENT

The characteristics of natural water catchments are shown in the following illustration (Figure 6.1).

In most countries we observe increased flooding. This can be partly related to climate change and more intense storms, but also due to a lack of attention to catchments with loss of wetlands, loss of woodlands to farming and earlier policies of 'straightening' rivers with loss of flow attenuation from meandering flow. Countries, including England (Rey *et al.*, 2026) are developing strategies to improve control of rainfall runoff (Figures 6.2 and 6.3).

There are good examples of how water catchments can be retrieved (Rivers are Life, n.d.). The 'Rivers are Life' reference covers examples in the Netherlands, France, the USA and Japan. The example from the Netherlands is reproduced below.

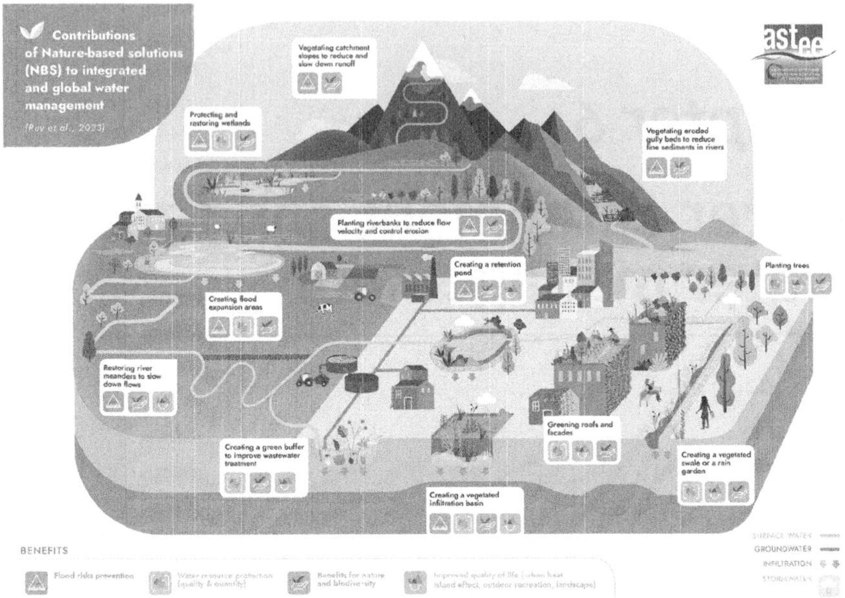

Figure 6.1 Nature-based solutions for water management with benefits for environment and society. It includes some elements of the urban aspects of Part 2 (Source: Rey *et al.*, 2026).

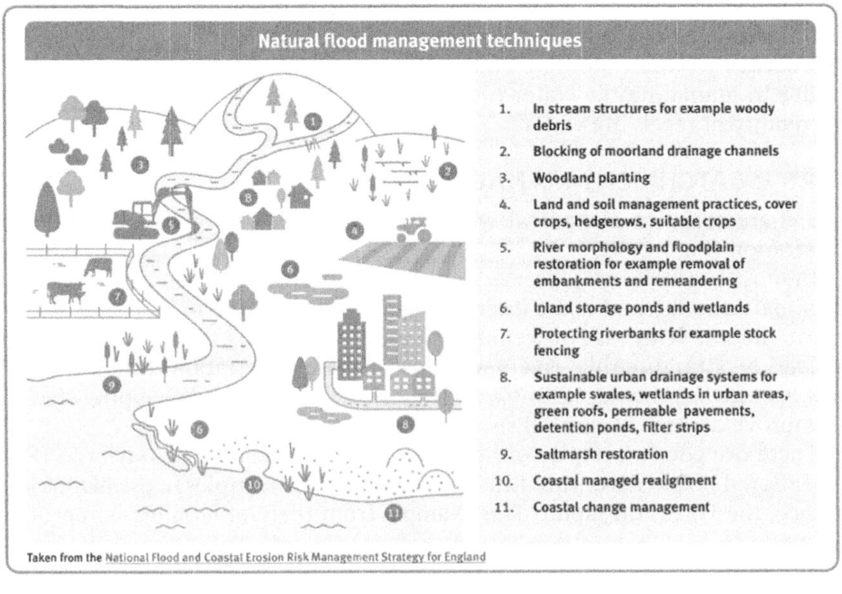

Figure 6.2 Flow attenuation (Environment Agency, 2022).

Figure 6.3 Example of a meandering river.

The Dommel River, Netherlands: nature-based solutions

In the Netherlands, the Dommel River restoration project showcases the effectiveness of nature-based solutions in improving river health and resilience. Historically, the Dommel River suffered from pollution, habitat degradation, and flooding exacerbated by urbanization and agricultural practices.

To address these challenges, a multifaceted restoration approach was implemented. This included:

- **Floodplain restoration:** By restoring natural floodplains and creating retention areas, the Dommel River's capacity to absorb and mitigate floodwaters has been enhanced. This reduces the risk of flooding for nearby communities and provides valuable wildlife habitat.
- **Water quality improvement:** Efforts to reduce pollution from agricultural runoff and urban sources have led to improvements in water quality along the Dommel River. This has benefited aquatic ecosystems and promoted recreational activities such as fishing and boating.
- **Biodiversity conservation:** Through habitat restoration initiatives, including the planting of native vegetation and the creation of fish passages, the Dommel River has become a haven for biodiversity. Endangered species such as the European otter have returned to the river, signalling the success of conservation efforts.

The Dommel River restoration project exemplifies the importance of integrating nature-based solutions into river management strategies. By working with, rather than against, natural processes, the Dommel River has been transformed into a resilient and thriving ecosystem.

Riverbank Rehabilitation Project, Australia (NSW Government, 2026)

Local Land Services is delivering this project to protect and improve flood-damaged riverbanks across 11 New South Wales catchments. The Riverbank Rehabilitation Project supports land managers to prepare for, respond to and recover from the impact of future flooding events and by identifying opportunities to improve the waterway approvals process.

The key activities of the Riverbank Rehabilitation Project are riverbank rehabilitation at targeted sites for bed and bank stabilisation in partnership with Soil Conservation Service, river experts giving land managers practical advice on caring for waterways and getting planning approvals and reviewing planning and approval barriers that make riverbank rehabilitation hard for land managers and recommend ways to improve the process.

Examples of projects:

Rehabilitating a site on the Wilson River, New South Wales and Queensland

The rehabilitation of a site on the Wilson River in the Hastings catchment highlights the work of the project. In collaboration with the Soil Conservation Service, rock and timber structures were installed along with planting 2030 native tube stock plants to ensure long-term protection of the riparian environment and prevent active erosion.

Rehabilitating a site on the Rous River, New South Wales

Nobbys Creek was close to breaking its banks and joining the larger Rous River. This would have reduced its length by 150 m, trigged upstream erosion due to a head cut and increased erosion around a newly isolated island. The Riverbank Rehabilitation Project delivered on-ground works including stock exclusion fencing, weed control measures and the planting of native riparian vegetation. See website (reference 3) for images on a section of the actively eroding riverbank before and after stabilisation works.

Nature-based Solutions (NbS) in Asia (Asian Disaster Preparedness Centre (ADPC) and International Union for Conservation of Nature (IUCN), 2025)

A World Bank-financed project covers catchment remedial work in four countries, India, Bangladesh, Pakistan and Nepal. The role of NbS is recognized in climate mitigation. India's national contribution includes several ecosystem-based mitigation measures to achieve a carbon sink of 2.5–3 billion tonnes of CO_2 equivalent through additional forest and tree cover by 2030. The recognition can be further strengthened through the inclusion of specific targets for wetlands, considering its role as a carbon sink. The policy promotes convergence and pooling of resources from national flagship programs such as Mahatma Gandhi National Rural Employment Guarantee Act (MNREGA), Watershed Development Program and Jal Jeevan Program. The role of ecological connectivity in achieving water security is recognised and includes conservation of river corridors, water bodies and infrastructure.

Wetlands

Wetlands are part of natural water catchment restorations given above, but they are so important they are highlighted here (UN Environment Programme, 2022). With increasing demand for food, only 15% remain of those in existence in the year 1700. Their importance as a haven for wildlife, in filtering pollution and as stores of carbon are now recognised, this providing much greater focus on their restoration by UNEP. See UN Decade on Ecosystem Restoration 2021–2030 (UN Decade on Ecosystem Restoration).

A good example is The Okavango Delta in Botswana (UNESCO World Heritage Convention). It is a UNESCO World Heritage Site. I reproduce the UNESCO description. 'This delta in north-west Botswana comprises permanent marshlands and seasonally flooded plains. It is one of the very few major interior delta systems that do not flow into a sea or ocean, with a wetland system that is almost intact. One of the unique characteristics of the site is that the annual flooding from the River Okavango occurs during the dry season, with the result that the native plants and animals have synchronized their biological cycles with these seasonal rains and floods. It is an exceptional example of the interaction between climatic, hydrological and biological processes. The Okavango Delta is home to some of the world's most endangered species of large mammal, such as the cheetah, white rhinoceros, black rhinoceros, African wild dog and lion'.

Atmospheric rivers (PBS News, 2025)

Sometimes called rivers in the sky, these are long channels of water vapour, typically carrying a flow the equivalent of twice the Amazon River. They are important sources of water in many areas, for example in Western US where they provide 30–50% of the annual precipitation. The 'rivers' travel at high altitude until they reach a barrier such as a mountain where the atmosphere is cooler (and no longer able to carry that amount of water) and they descend into intense storms. The concern now is that with higher atmospheric temperatures they carry more water and are more likely to cause flooding. There is also an increase in their frequency with subsequent storms falling on saturated ground.

PART 2: URBAN WATER MANAGEMENT

Over and above river controls achieved in the natural catchment, controlling rainfall run-off in the cities is necessary. There are a number of aspects:

- Wastewater – Legislation and Standards
- Separate or combined sewer systems and storm overflows
- Managing surface water runoff
- Flood defences

Wastewater treatment – quality discharge standards

An important part of wastewater management history is the British Governments' establishment of a Royal Commission in 1898 on treatment and disposal of sewage. The Commission produced a number of reports with a report in 1912 which established the '20/30 standard' for disposal of sewage to rivers and coastal waters. The 20 mg/L for biological oxygen demand (BOD) and 30 mg/L for suspended solids (SS) limits were adopted. They became the 'de facto standard' in many parts of the world.

The EU Wastewater Treatment Directive 1991 (European Union, 2024) set standards for BOD, chemical oxygen demand (COD), SS, total phosphorus and total nitrogen. The revised EU Wastewater Treatment Directive 2024, relates to a wider range of treatment works and goes much further, covering energy, polluter pays, and monitoring for antimicrobial resistance, viruses such as Covid, and microplastics and PFAS (often referred to as forever chemicals). It promotes greater wastewater re-use for large cities. Member States will be required to develop integrated management plans to deal with stormwater to reduce the risk of urban floods and pollution during heavy rain. The Cuncliffe Regulation Report recommends that it becomes the regulatory requirement to be enforced in England. It will be interesting to see how implementation is handled throughout Europe and what impact it might have across the world.

Separate or combined sewer systems and stormwater overflows

It is desirable to have separate sewer systems so that surface water drainage does not become contaminated by sewage and can flow into the river or the sea without degrading the water quality. Successful separate systems require good documentation and careful regulation. Without controls developers and plumbers make uninformed connections with the result of the system carrying sewage becoming a combined sewer, and sometimes the surface water drains also carrying sewage. Combined systems are designed so that in severe storms they overflow to avoid overwhelming the sewage treatment works. This should not damage river quality as the discharges are highly diluted and flow into high volume river flows. If there is a lack of investment in storm overflows to cater with a growing population and expanding building, storm overflows operate frequently and create poor river quality or sea and beach pollution. This was the situation in parts of England and is described in chapter 11.

Managing surface water runoff

It is necessary to control rainfall runoff from roads and buildings. Alongside the large roads entering cities there should be soakaways called swales. See examples in Figure 6.4.

Also, runoff from roofs should flow into soakaways or be collected for water storage in water butts or underground tanks. Surfaces around buildings should be porous. Roof gardens slow down runoff and provide valuable additional plant growing areas.

Reading the summer edition of the National Trust Magazine (National Trust Magazine, 2026), I was delighted to see the success that the Trust's gardeners with volunteers have achieved in floods and drought management, for example run-off from greenhouse roofs being collected, underground storage tanks, natural flood management schemes, building leaky dams and ponds. I like the term 'Slow the Flow'. All such experience needs to be shared. These local activities need to be applied worldwide across the scale of gardens to whole water catchments.

Figure 6.4 Swales.

Flood defences

The investment priority should be to reduce run-off, as discussed in Part 1. At the same time construction of buildings in flood plains should not be allowed. Flood plains are just that, areas where excess water from full rivers and streams can be stored until the river or streams reduce and the flood water can return and be carried on safely. However, it is unlikely that flood control provisions can be made to cope with the most extreme storms. Early warning systems are vital, either for evacuation measures when life is endangered or to set-up flood defences (see chapter 9).

So, to what extent can houses, other buildings and infrastructure be made flood resistant? There is the general claim that cost of effective defences greatly exceeds damage and cleanup costs, but that does not take into consideration the emotional impacts.

A UK Government report (UK Government, 2025) concludes that defences up to 600 mm can be effective if they are bespoke designed, professionally fitted, and well maintained and stored.

US company Dam Easy Flood Barriers' article, 'Myth vs Truth About Flood Proofing in 2025' (Dam Easy Flood Barriers, 2025) relates to purchasable products but does give valuable advice.

PART 3: WATER STORAGE

Although estimates vary, aquifers make up 99% of available freshwater. There is 60 times as much groundwater than that in all lakes and rivers combined. However, water from many aquifers has been over extracted and the aquifers won't recover naturally in human time frames. Recovery will require artificial recharge and where would that water come from? Building large reservoirs by damming rivers not only takes water from users downstream but has limited life, due to droughts where there is insufficient rainfall to maintain levels to make up for evaporation losses. Droughts cause hydropower schemes to become less effective when reservoir levels fall, leaving less water to drive the turbines (World Economic Forum, 2023). This is reducing the amount of investment in hydropower.

So, what to do? The World Bank (World Bank Group, 2023) advocates an integrated approach. Their 2023 Report titled *What the Future Has in Store: A New Paradigm for Water Storage* is 'an urgent appeal to practitioners at every level, both public and private, and across sectors, to come together to champion integrated water storage solutions – natural, built, and hybrid – to meet a range of human, economic, and environmental needs for the 21st century. Closing storage gaps will require a spectrum of economic sectors and stakeholders to develop and drive multi-sectoral solutions. The proposed integrated water storage planning framework is grounded in sustainable development and climate resilience, with the potential to pay dividends for people, economies, and environments for generations.'

It seems to me that effective action is more likely to be taken at local levels with communities working together. Such areas are more likely to experience common weather conditions. Also, the lead can be taken by those with consistent managements systems. One example is of a city lead in the USA (Jones, 2023). Another aspect is the use of local storage vessels, storage tanks and towers (McClure, 2025). A more integrated approach is Strategic Water Management with Community Involvement and Advocacy (Sustainability Directory, 2025). An effectiveness study based on examples in Tigray, Ethiopia and Rajasthan, India (Carter, 2025) demonstrated that community involvement achieved 30% more water conservation than top-down projects alone. The community projects used a combination of rainwater harvesting and xeriscaping. There was the added benefit of achieving greater community cohesion.

TRANSITION ARRANGEMENTS

Generally, current activities to control run-off from the natural catchment and those of urban water management are separate both in management and financing. Focussing on coping in a water bankruptcy era will require an integrated approach. This could require city-based operations with smaller water service delivery units. The requirements should be considered in early planning stages.

REFERENCES

Asian Disaster Preparedness Centre (ADPC) and International Union for Conservation of Nature (IUCN). (2025). Enhancing Water Sector Resilience Through Nature-based Solutions in South Asia: Key challenges and opportunities, https://iucn.org/sites/default/files/2025-08/regional-guidance_nbs_water_sector_south-asia.pdf

Carter N. (2025). Case Studies: Successful Community-Driven Water Conservation Efforts, H20Conserve.org, https://h20conserve.org/case-studies-successful-community-driven-water-conservation-efforts/

Dam Easy Flood Barriers. (2025). Flood Defence Myths vs. Brutal Truth. And why flood barriers for doors after all, https://dameasyfloodbarriers.com/a/blog/fllod-barriers-for-doors-myths-vs-truth

Environment Agency. (2022). National flood and coastal erosion risk management strategy for England: executive summary, https://www.gov.uk/government/publications/national-flood-and-coastal-erosion-risk-management-strategy-for-england--2/national-flood-and-coastal-erosion-risk-management-strategy-for-england-executive-summary

European Union. (2024). EU Wastewater Treatment Directive 2024/3019, https://eur-lex.europa.eu/eli/dir/2024/3019/oj/eng

Jones P. S. (2023). Investing in Local Water Infrastructure for a Resilient, Livable Future. National League of Cities, (NLC), https://www.nlc.org/article/2023/04/04/investing-in-local-water-infrastructure-for-a-resilient-livable-future/

McClure. (2025). Planning for Water Storage Tanks, Towers, and Beyond, https://mcclurevision.com/planning-for-water-storage-tanks-towers-and-beyond/

National Trust Magazine. (2026). Summer 2026. Rooting For Nature, https://magazine@nationaltrust.org.uk

NSW Government. (2026). Riverbank Rehabilitation Project, https://www.nsw.gov.au/regional-and-primary-industries/natural-resource-management/protecting-waterways-on-farms/managing-riverbank-erosion#toc-on-ground-success-stories

PBS News. (2025). What's an atmospheric river? AP explains the weather phenomenon, 23 December 2025, https://www.pbs.org/newshour/science/whats-an-atmospheric-river-ap-explains-the-weather-phenomenon

Rey F., Bertrand-Krajewski J.-L., Fernandez S., Guerrin J., Herivaux C., Lafforgue M., Le Coent P., Pons M-N. and Rulleau B. (2026). Nature-based solutions for water management: pluridisciplinary state-of-the-art and research needs. *Ecological Engineering*, **225**, 107887, https://doi.org/10.1016/j.ecoleng.2025.107887

Rivers are Life. (n.d.). River Restoration Success Stories: Lessons Learned and Best Practices, https://www.riversarelife.com/articles/river-restoration-success-stories-lessons-learned-and-best-practices

Sustainability Directory. (2025). Community-Based Water Management and Equity, https://prism.sustainability-directory.com/scenario/community-based-water-management-and-equity/

UK Government. (2025). FloodReady – an action plan to build the resilience of people and properties, https://www.gov.uk/government/publications/floodproof-an-action-plan-to-build-resilience/floodready-an-action-plan-to-build-the-resilience-of-people-and-properties

UN Decade on Ecosystem Restoration. Preventing, halting and reversing loss of nature, https://www.decadeonrestoration.org/

UN Environment Programme. (2022). Wetlands: the unsung heroes of the planet, https://www.unep.org/news-and-stories/story/wetlands-unsung-heroes-planet

UNESCO World Heritage Convention. Okavango Delta, https://whc.unesco.org/en/list/1432

World Bank Group. (2023). What the Future Has in Store: A New Paradigm for Water Storage, https://www.worldbank.org/en/topic/water/publication/what-the-future-has-in-store-a-new-paradigm-for-water-storage

World Economic Forum. (2023). Hydropower: How droughts are affecting the world's biggest renewable energy source, https://www.weforum.org/stories/2023/11/hydroelectricity-generation-falls-droughts-climate-change/

doi: 10.2166/9781789065527_0059

Chapter 7

Communications and public participation
It's good to talk

COMMUNICATIONS

This chapter is vitally important in identifying communications on the key changes and actions necessary to address key water-related issues listed below.

(1) Understanding water bankruptcy and the need for change.
(2) The need for governments to recognise the importance of demand management in community water supply systems, particularly the need for water distribution leakage control as the largest source of 'new' water and for sustainability.
(3) Recognising that water for food production is no longer secure. Understanding the issue of 'virtual water' and the consequences for producer countries on efficiency, crop selection and methods of application.
(4) Water storage must be increased.
(5) Information to the public on early warning of floods is critical so that they can take the necessary evacuation and other safety actions. The lessons learnt from each warning need to be fed back into the early warning models.
(6) Early warning systems for droughts are different than those for floods as actions need to be taken from the precautionary through to day-to-day initiatives as necessary.
(7) The public needs to be informed that investment in water including catchments, needs to be increased significantly.
(8) Penalties on water service providers for failure to meet requirements should be replaced by penalties imposed on managements.
(9) The need for government sponsored community groups to facilitate changes.

Understanding 'water bankruptcy (Madani, 2026) and the need for change'

The introduction to chapter 1 describes the background to the study of the UN University in Canada and why the approach to water matters needs to change. It explains that even without climate change the world population's water requirements now exceeds the rainfall necessary to provide for the current water supply conditions. It is no longer a case of expecting to be able to cope with crises by trying harder. We are in a post-crisis era.

In practice the impact will be mostly in the developing world with its rapid population growth and more arid climates. However, with climate change, anywhere in the world could suffer a severe drought as in the case of Cape Town, South Africa in 2015 (Smedley).

The 'how' in most of the chapters in this book will not change but become even more important. The big changes required are in government policies as presented in chapter 1, and (in this chapter) how those changes are communicated.

The need for governments to recognise the importance of demand management in community water supply systems

Generally, although some governments have put in place investment to reduce water demand, there has been a tendency to put that in 'the too difficult box' on the understanding that it can wait until tomorrow or that additional new water resources could be developed. In many cases such new resources have been vanity projects with the opportunity of putting a plaque on the wall and having an opening ceremony. The better option would have been to introduce extensive water distribution system leakage reduction programmes. In this post-crisis era the only option is to reduce leakage as the only 'new' water resource which is available. Just reducing leakage from 50% to 30% provides 40% more water. There is a full discussion of reducing demand, and thus increasing supply is given in chapter 3.

Recognising that water for food production is no longer secure

Quoting from UN Report: Agriculture accounts for over 70% of global freshwater withdrawals, providing food, employment opportunities, and stable livelihoods to billions of people. But about 3 billion people and over half of the world's food production are located in regions that are already experiencing, or are projected to face, declining trends in total water storage. Around 1.2 billion people already live in agricultural areas facing severe water constraints, and over 170 million hectares of irrigated cropland are under high or very high water stress.

The problem is compounded by 'Virtual Water' in crop producer countries, typically lower income countries whose economies rely on providing produce which cannot be grown in many wealthier parts of the world, such as Europe and North America. The classic item quoted is coffee which requires 14 litres of water to produce one cup of coffee. Often there is insufficient rainfall to provide for such agriculture with the consequence of over pumping of aquifers perhaps to the point of non-recovery, see chapter 4 on irrigation. Buyers, such as Supermarkets need to be involved in solutions together with businesses with the ability to develop crops requiring less water and crops which will grow in colder climates. The current producer countries will require financial support to compensate for loss of business.

WATER STORAGE MUST BE INCREASED

In particular, this applies to aquifers which are able to be recharged but also to reservoirs and ponds. The water used for 'new' storage should not come from existing sources but from rainwater. This as part of urban water management is discussed in chapter 6.

Information to the public on early warning of floods

This is increasingly important as is the need for governments to recognise the need for investment in managing rainfall run-off and in using the information provided by the public from actual flood events in improving early warning systems: see chapter 8.

Early warning systems for droughts

There can be a tendency to think it won't happen to us. The experience of Cape Town (Smedley) should be a wakeup call that it can happen anywhere. Governments need to put in place plans for gradually reducing the amount of water available and for the public to know what reducing water availability would mean in practice: see chapter 9.

The public needs to be informed that investment in water, including on catchments, needs to be increased significantly

At the time of writing the world was in political and economic turmoil, with the prospects of a global recession. Governments had to decide which water saving measures should be given priority, such as directing farmers to grow crops most suited to local conditions.

Penalties on water service providers for failure to meet requirements should be replaced by penalties imposed on managements

It is to be expected that water customers angry with poor performance like to see water service providers penalised. The problem is that fining the service provider takes away money from the important investment that needs to be made. The solution is to target those individuals responsible for the failures.

The need for government-sponsored community groups to facilitate changes

Water bankruptcy is going to affect all communities and their involvement is going to be important. This leads naturally to public participation.

PUBLIC PARTICIPATION

Levels of public participation

Levels of public participation were very effectively defined in 1969 by the Arnstein ladder (Table 7.1).

Table 7.1 Arnstein's ladder with brief descriptions of each rung (source: author, taken from information in Arnstein, 1969).

Level	Form of Participation	Description of Participation
8	Citizen control	Examples given by Arnstein are funding of communities to run their own co-operatives and development projects
7	Delegated power	Citizens have dominant decision-making authority on a specific item or plan, through a majority of seats
6	Partnership	Power is redistributed through negotiation between citizens and power holders for example joint policy boards
5	Placation	A token degree of influence by having hand-picked individuals on Public Bodies
4	Consultation	A further step (public meetings) but offers no assurance that concerns or ideas will be taken into account
3	Informing	An important step but tends to be one way with no channel for feedback
2	Therapy	Citizens who may be complaining are given therapy to divert their attention – not relevant to water but healthcare
1	Manipulation	Citizens put on advisory committees for the express purpose of educating them or engineering their support

There have been many variations of the ladder but, by and large, Shelley Arnstein's structure has remained a useful classification.

The situation in 2026

Over 50 years later, almost any paper on public participation includes reference to Arnstein's ladder. Humphreys (2002) refers to 'its continuing usefulness and relevance'. In the developed world the two bottom rungs of the ladder tend now to be discarded, as it is generally felt that governments are now serious about public participation. Arnstein had labelled levels 1 and 2 as nonparticipation (Figure 7.1). The question now in the developed world is the extent to which citizens are given power over local plans. In the developing world, with less experience of public participation, it is important that governments avoid the temptation to manipulate citizens by issuing incomplete and biased information. Real and effective public participation may be seen by public officials as diminishing their power, rather than a means of strengthening their authority achieved through gains in public support and in information on citizens' requirements.

A much simpler framework of three levels – consultation, presence and influence – is given by Goetz and Gaventa (Gaventa & Goetz, 2001). OECD (2001) also give three categories, those of information, consultation and active participation. See also the planning and management model set out in GIZ Databank on Rural Water Supply, 2024.

Remarkably, the Shelley Arnstein ladder is still discussed today. An organisation which advises UK local government discussed Arnstein's Ladder and its value

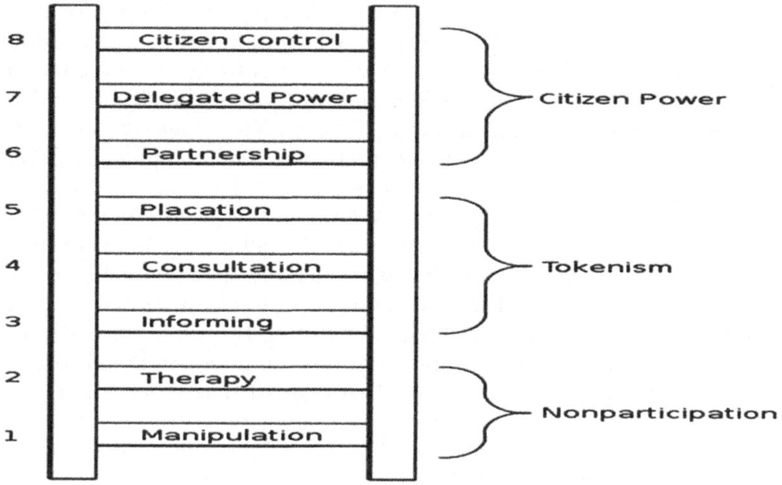

Figure 7.1 Arnstein's ladder (source: Shelley Arnstein (Arnstein, 1969)).

(Zencity, 2023). It concluded that the two highest levels, namely Citizens Control and Delegated Power, were not appropriate for local government. Based on its experience it has developed An Engagement Framework under the headings of Accessibility, Transparency and Timing and Process. Accessibility involves the use of various channels tailored to the organisation involved. Transparency is seen as making sufficient information readily available with the scope of influence publicly stated right from the beginning in accessible language. Timing and Process related to a statement of community need in an iterative process to satisfy the citizens.

THE ROLE OF GOVERNMENTS

Effective public participation in developing countries depends critically on government policy in establishing a framework in which there are sound overall policies. What can people do to ensure that such policies result in their obtaining the necessary water and sanitation services? WaterAid, an established UK NGO with extensive experience in developing countries, is pursuing a programme of transformational change (UN Department of Economic & Social Affairs, Sustainable Development, n.d.).

Examples of public participation

Accra, Ghana

Some people living on the outskirts of the city of Accra in Ghana do not have access to a continuous supply of tap water. In some areas they are wholly reliant on tanker and cart supplies, and in other areas the tap supplies are intermittent. Local Community Groups have formed to seek improvements in their water supply situation. One such group is the Teshie Concerned Citizens Association, who say that they have not received a regular tap water supply for 20 years. At the end of December 2002, the Teshie Association held a press conference with an appeal to the Ministry responsible for water to restore a regular supply. In February 2003 members of the Association took to the streets in Accra to protest at what they called 'the failure of Ghana Water Company Ltd to restore and maintain a regular supply of water to Teshie'. The company response was that one of the main reasons for the shortages was the amount of water taken through illegal connections made by residents in the area. The Greater Accra Regional Minister set up a committee, which included a member from the Association, to investigate the Teshie water supply situation. The committee, confirming the seriousness of the situation, reported in June 2003. The report identified eight causes of the problems, including the rapid expansion of the area, and made a number of recommendations. One recommendation was the completion of the development of local storage so that more water could be taken during the times when distribution system pressure was high enough to supply the Teshie area. The Association has been successful in obtaining political focus on their problem (Ghanaian Times, 2019).

India – Alwar District

There are many good examples of effective public participation on a large scale in India, often in rainwater harvesting schemes, in which local action is taken to develop new systems or to rehabilitate systems which had been in operation previously for hundreds of 3 years. One well-documented example (Community-led Water Management in the Alwar District, Rajasthan, India, n.d.; Tarun Bharat Sangh, Rivers Rejuvenated, Arvari River, n.d.; Rapid Transition Alliance, 2020) is in the north Rajasthan district of Alwar which since 1985 has been transformed through local community effort, supported by the NGO Tarun Bharat Sangh (TBS). Following a severe drought in 1985, a local community leader suggested to TBS that the poor water resource situation could be alleviated through a traditional rainwater harvesting approach. This involved the construction and rehabilitation of earthen check-dams, which in India have various names, such as johads (Figure 7.2), khadins and ahar pynes, depending on the topography, the type of soil and the details of construction. The check-dams hold back water during the rains so that groundwater is recharged, or the soil is irrigated for agriculture. The work started in a small way in pioneering villages, but grew rapidly when neighbouring villages observed the benefits of the johads. Since 1985 more than 4500 johads have been built in over 1000 villages raising the level of the water table in the area. As a result, the Arvari River, which had been dry for many years, returned to full flow in 1996. This has transformed the economy of the area with growth of profitable agriculture and fishing. The construction of johads was carried out largely with local labour but with some hired skilled labour. Also some materials not available in the village had to be purchased. TBS acted as a facilitator and advised on setting up the village assemblies, called Jal Samitis, to achieve strong community involvement in decision making. It also assisted the villages in obtaining external funding, and established the principle that the communities should pay for at least one quarter of the cost. Later this was increased to one third of the cost. Villages

Figure 7.2 An example of a johad (check-dam) (source: Centre for Science and Environment, India).

established funds to pay for maintenance of the johads with each family being required to contribute 2.5% of its income to the fund, or to provide one day of voluntary effort per month.

The individual local management of the water resources has developed into a local water parliament. The re-establishment of the river to normal and continuous flow conditions resulted in the return of fish to the river, and the Rajasthan Government let a contract for commercial fishing in the river. Although the villagers were vegetarian, and not interested in the fish, they saw this move as a possible step by government to take over control of their water resources. So in 1999, the communities formed the Arvari Parliament to manage the Arvari River and its tributaries. The Parliament, which has no legal status, has 142 members who are nominated by the 72 village assemblies. This body has established rules for the use of the rivers, and has a coordinating committee to handle operations and enforce the rules, albeit on a voluntary basis. The rules include those related to extraction for irrigation, land use and ownership, and protection from industry. In practice, the Parliament is working well with effective resolution of disputes and is safeguarding the water resources through preventing the growth of water-using industries.

The success of the Alwar initiative has stimulated international agencies, such as the United Nations Development Programme, and the Government of Rajasthan to support the Jal Bhagirathi Foundation (DFID, 1999), which was formed in 2001, to establish water conservation measures in the Marwar belt of the State. TBS is providing advice, including that on setting up local committees, the Jal Samitis, to include women and representatives of the weaker parts of the communities. Since 2001 the Foundation has been very active with support to communities in the construction of over 70 storage units serving over 14 000 people. The Foundation signed an agreement in 2003 with UNDP for a project on 'vulnerability reduction through community management and control of water in drought-prone areas of the Marwar region'. Another agreement was signed with UNICEF for a project on 'community empowerment for drought management'. It will be interesting to see the extent to which this concerted effort to expand community-managed johad schemes will yield sustainable water supply systems over the next 20 years.

India – Kolhapur

This example involves a scheme introduced in 1980 for the supply of river water to four villages in the Kolhapur District in Maharashtra. This system was the subject in a Water and Sanitation Program Field Note (Hasan, 2024). The policy was for districts to take responsibility for the operation and maintenance for such systems, but in this case the Kolhapur District did not have the financial resources to do so. The Maharashtra Government persuaded the local administration of the four villages to take on the operation and maintenance of the system. In 1981, the villages formed a joint water management organisation, called a Mandal in India, to operate and maintain the system on an informal basis. Initially there were 419 house connections but, through installing additional pipelines and encouraging more people to connect to the system, by 1999 the number had grown to 1723. Those with house connections paid an initial connection charge, an annual water charge and an annual development charge as part of financing system enhancements. Those villagers without a house connection are able to take water free of charge from public standposts, with each village making an annual financial contribution to the Mandal for the maintenance of this public facility. Over the period 1981–1999, the Mandal was able to achieve a revenue surplus. The accounts are audited each year and are available for public scrutiny. The system is not without its problems, with deterioration of the river water quality and the need for replacement of some of the expensive equipment, such as the pumps, but its overall performance greatly exceeds that of neighbouring systems managed by the District. For example, because it has been necessary to achieve financial stability, the Mandal introduced a policy of disconnection to ensure that householders pay their bills. This was possible without accusations of penalising the poor due to the availability of public standposts. However, by 1999 only 16 disconnections had been necessary. In comparison, the other District managed systems have no such policy with a much higher level of non-payment and the risk of unsustainable systems. This example demonstrates that self-management generates a high level of community responsibility, and that generally nonpayment is not a case of people not being able to pay for water services, but not wanting to pay for them.

Orangi, Karachi, Pakistan

One of the largest and most-cited examples of successful community participation and action is the Orangi Pilot Project, OPP. Orangi is a township west of Karachi city with a population of over 1 million people. The project (Israel, 2009) was established in 1980 by an eminent social scientist, Akhtar Hameed Khan, who believed that government programmes were failing because government officials and their professional advisers did not have a sufficiently good understanding of the culture, the sociology, or the economics of low-income communities, especially those in informal settlements. His research led to the identification of four important needs: those of sanitation and housing, employment, health and education. Within that group of important requirements, sanitation was seen by the Orangi residents as of the highest priority. Four NGOs were formed with the Orangi Pilot Project – Research and Training Institute (OPP-RTI), covering sanitation, housing, education, research and training.

The approach taken was for OPP-RTI to hold meetings with those residents willing to establish a lane (street) Organisation and appoint a lane manager. OPP-RTI offered technical assistance to lanes to build their own sewerage systems. Typically, a lane consists of 20–40 houses. Initially, lanes with a natural drainage channel would build their latrines and sewer drains, and later this was expanded to include connector sewers to take sewage from other lanes. There has been criticism of the use of natural drainage channels, but initially this was necessary for affordability reasons and because of the limit of community influence. The longer-term intention was for the Karachi Corporation to convert the channels into box trunk sewers leading to sewage treatment works. By 2001, over 6000 lanes with a total of over 90 000 houses had built their sewerage systems. Due to choice of technology, self-construction and self-financing, the cost per house was only US$16.5, and therefore affordable.

As a result of the success of OPP, the approach has been replicated in other areas, some through local government initiatives, and some with financial support from outside agencies. Some have succeeded, some have failed. Arif Hassan, in his paper, gives the reasons for the successes as follows:

- a combined technical/social organisation which included members of the community;
- a leader who could establish an effective relationship with local government;
- the existence of a good map of the area;
- patience;
- availability of funding for collector sewers;
- advice and training from OPP-RTI;
- effective monitoring and management through regular weekly project meetings;
- transparency of accounts and involvement of local people on the board of the NGO involved.

He gives the reasons for failures as the lack of these above items, plus acceptance of large sums of donor money leading to financial mismanagement, and subsidising of lane developments which had resulted in disputes, higher costs and loss of empowerment of the communities.

The challenge for OPP-RTI, given its resource limitations, has been how to meet the demand for its expertise and experience. It is attempting to meet this through training other NGOs and running lecture series for communities and elected officials. It is proving difficult to replicate the understanding and necessary culture that developed within OPP-RTI which is critical in providing communities with the required advice for success.

San Luis, Agusan del Sur, The Philippines

The training necessary for communities to manage water and sanitation systems in the area of Agusan del Sur in the Philippines has been provided by an NGO, The Philippine Center for Water and Sanitation, through the establishment of a

Provincial Water and Sanitation Center. The physical water supply system was provided by the local government for community management. The Center's role was to build the capacity of municipal level professionals to support the community water and sanitation associations. An example of the success of this scheme has been in the community of Barangay Doña Flavia in the Municipality of San Luis (Israel, 2009), where system coverage has been developed to serve 100% of the population of 2687 people. Importantly, the community receives a constant 24 hour a day supply. The approach taken was for the Center to hold orientation seminars for community organisers and to train trainers who would assist the community teams. The training included both the concepts of sustainable systems, and the more detailed aspects related to hygiene promotion, technology and tariff calculations. The success in San Luis has stimulated interest in the approach elsewhere in the Philippines, with 33 municipalities in the south of the country in Mindanao pursuing community managed water and sanitation systems.

Effective public participation

The examples given above show the value and importance of public participation in the provision of water services, without which there cannot be understanding and support for necessary policies and actions. The examples also show that the level and type of participation is dependent on whether or not there is already a good water service, and dependent also on the size of the community and the complexity of the water service operations. Crucially, it is dependent upon whether or not there is already a degree of transparency. Therefore, I draw my conclusions first on public participation in developed countries, and then in the developing countries making distinctions between urban and rural communities.

Developed countries

Generally, in most of Western Europe, much of North America, and in countries like Japan, water services are well-developed, and the general public are not clamouring for direct control of them. Most people do not take much interest until they receive a bill or something goes wrong. They expect to be able to turn on the tap and receive clean water, and flush the toilet without thinking about how that water service is managed. Equally, the media take little interest when everything is working well. The situation changes when there are drought conditions, and water usage is restricted, or when there is a pollution incident. But by and large, 'out of sight and out of mind' applies most of the time. On the other hand, the public expects to receive accurate and timely information to give reassurance on drinking water quality and on protection of the environment. The public become very concerned if they suspect that something is being hidden, or if they feel that they are paying excess charges for their service. The latter becomes an issue in situations where the water service provider is from the private sector. They do not like to see companies making profits at a time when they have a restricted water service, not recognising that profits are necessary to pay for investment, and are the driver for greater efficiency. This important issue is discussed in chapter 11 but is mentioned here because of its relevance to public attitudes and understanding. The public appear to believe that good reliable independent information and transparency are even more important when private companies are involved. I believe that there is an equal requirement for transparency with public sector operations, as with less separation of policy and delivery, there are greater opportunities for hiding inefficient operations. So whether with public or private delivery, I see consumer councils as a necessary part of achieving transparency, as most members of the general public have neither the time nor the knowledge to be able to scrutinise published data. There is a need for a dedicated independent professional body to act on the public's behalf, and I see the UK Consumer Council for Water as a highly effective model.

Developing countries – rural areas

In rural areas, with relatively small communities compared with urban situations, as shown by the examples given in this chapter, it is possible for people to be involved directly in decisions on the provision of water services, and for communities to be organised so that they can provide much of the labour in the construction and operation of those services. There are many successes, often through the facilitation of NGOs who provide guidance and who assist in obtaining financial support. History suggests that direct replication of effective examples may not work, as success is highly dependent on local conditions. However, based on examples with positive results, I find the following are ingredients for success:

(1) Government support for local initiatives, as some examples in India show that, with NGO help, it is possible to make some initial progress without government support and, as a result, stimulate government to take an interest. Part of government's policy role is to provide the necessary legal framework to give communities security of tenure on their systems.

(2) Strong and effective community leadership to recognise the need for action, and with the vision to see what assistance is required, as in the case of the Alwar district in India.

(3) A respected NGO with the knowledge to facilitate projects, provide training, and give policy, technical and financial guidance.

(4) Choice of 'technology' which allows for local production and maintenance, and which provides not only the required quantity but a microbiologically safe drinking water.

(5) A policy of charging for water, leading to full-cost recovery for sustainability, but with provision for the poor as in Kolhapur in India. Full-cost recovery, with financial provision for refurbishment of systems is essential for long-term sustainability, and it also engenders a sense of ownership and responsibility. It is essential that attention is not given to those well-meaning people who promote the view that the poor cannot afford to pay for water. Those poor, without an official supply, are having to afford much higher charges through purchase from water vendors. What the poor cannot do is save to pay monthly bills, so provision for them to 'pay as they go' is needed. This is discussed in chapter 2.

Developing countries – urban area

There are two difficulties in urban areas. One is the large size of communities making it much more difficult to establish coherent community groups, a problem that is compounded in peri-urban areas in which communities have grown rapidly and haphazardly, without the benefit of established community spirit and values. The other difficulty, in terms of water supply, is the practical one of the lack of availability of suitable water sources and insufficient land for large-scale rainwater harvesting. This places much greater reliance on outside government or other institutional involvement in the provision of the water. The Orangi Pilot Project has shown what can be achieved by operating with small street units recognised by local communities. Clearly this requires much NGO effort and a lot of time. The lessons learnt include the need for long-term and consistent advice, patience and transparency. The challenge is how to build on the Orangi success and provide the impetus for communities to develop and operate systems on the scale necessary to achieve the Millennium Development Goal of halving the number of people without dignified sanitation by the year 2015.

The best is yet to come – sadly not

In drafting my first book in 2006 I was excited by the prospects of wide and accurate communications through social media. Sadly, the systems have been captured and used for misinformation. There are now multiple groups with distinct and opposing messages, so the dream of improved communications has been lost.

Transition arrangements

The most important requirement will be to inform the public about water bankruptcy and what it is likely to mean in changes to water supply and how they will be informed of developments and on the formation of community groups.

REFERENCES

Arnstein S. R. (1969) A ladder of citizen participation. *Journal of the American Institute of Planners*, **35**(4), 216–224, https://doi.org/10.1080/01944366908977225

Community-led Water Management in the Alwar District, Rajasthan, India. (n.d.). https://www.scribd.com/document/53410470/Water-Management-Case-Study

DFID. (1999). Sustainable Community Management of a Multi-Village Water Supply Scheme in Kolhapur, Maharashtra, India. Water and Sanitation Program Field Note SPI Series 2, https://www.ircwash.org/sites/default/files/WSP-1999-Sustainable.pdf

Gaventa J. and Goetz A. M. (2001). Bringing Citizens Voice and Client Focus Into Service Delivery. Institute of Development Studies, Working Papers, UK, 138.

Ghanaian Times. (2019). Teshie residents raise alarm over desalination plant, https://ghanaiantimes.com.gh/teshie-residents-raise-alarm-over-desalination-plant/

GIZ Databank on Rural Water Supply. (2024), published by Deutsche Gesellschaft für Internationale Zusammenarbeit (GIZ) GmbH, Germany, https://www.giz.de/sites/default/files/media/pkb-document/2026-02/giz2024-en-publication-technical-brief-water-supply-luapula-province-clean.pdf

Hasan A. (2024). Delivering low-cost, high quality sanitation – at scale, International Institute for Environment and Development, https://www.iied.org/delivering-low-cost-high-quality-sanitation-scale

Humphreys P.C. (2002). Effective Consultation with the External Customer. Institute of Public Administration, Dublin, Ireland. ISBN 1902448812, https://www.researchgate.net/publication/278006074_Effective_Consultation_With_The_External_Customer

Israel D. C. (2009). An assessment of the local service delivery of potable water in Dumaguete City and selected areas of Agusan del Sur Province. *Philippine Journal of Development*, **36**(2), 97–100, https://pidswebs.pids.gov.ph/CDN/PUBLICATIONS/pidspjd09-2water.pdf

OECD. (2001). Citizens as Partners: Information, Consultation and Public Participation in Policy-Making, OECD Policy Brief No. 10, https://www.oecd.org/content/dam/oecd/en/publications/reports/2001/10/citizens-as-partners_g1gh14ab/9789264195561-en.pdf

Rapid Transition Alliance. (2020). https://rapidtransition.org/stories/nature-and-local-democracy-how-a-river-parliament-shows-what-community-control-can-do/

Smedley T. The Last Drop: Solving the World's Water Crisis, Picador, UK. ISBN 9781529058178.

Tarun Bharat Sangh, Rivers Rejuvenated, Arvari River. (n.d.). https://tarunbharatsangh.in/rivers-rejuvenated/#Arvari

UN Department of Economic and Social Affairs, Sustainable Development. (n.d.). WaterAid's agenda for transformational change, https://sdgs.un.org/partnerships/wateraids-agenda-transformational-change

UNU-INWEH Report, Madani K. (2026). Global Water Bankruptcy: Living Beyond Our Hydrological Means in the Post-Crisis Era. United Nations University Institute for Water, Environment and Health (UNU-INWEH), Richmond Hill, Ontario, Canada, https://doi.org/10.53328/INR26KAM001

Zencity. (2023). Arnstein's Ladder of Citizen Participation explained, https://zencity.io/uk/uk-blog-arnsteins-ladder-of-citizens-participation-explained/

doi: 10.2166/9781789065527_0073

Chapter 8

Water and wastewater treatment

INTRODUCTION

The main focus of this book is the management of water through capturing more water and reducing water demand. However, the water has to be safe and fit for purpose which requires water treatment. Equally, wastewater needs to be treated to protect the environment and to provide for source water for all purposes. The basic treatments are not discussed, but Puertos (2025) and Shook (2024) provide good general references.

The chapter covers the most important issues relevant to this book

(1) Antibiotics in wastewater
(2) Forever chemicals
(3) Plastic fibres in water
(4) Treated wastewater for potable use, for irrigation and for industry
(5) Desalination

Note: Water quality regulation is not covered, but the key aspects of chapter 4, Drinking Water Quality and Regulation in my book *Institutional Governance and Regulation of Water Services* are reproduced in the Appendix of this book.

ANTIBIOTICS IN WASTEWATER

The development of antibiotics has produced enormous benefits in people's health, but improper disposal has caused them to accumulate in the water environment. As a result, they now occur in source water needing to be treated for drinking water. Continual exposure to sub-inhibitory concentrations of antibiotics pave way to bacterial resistance to numerous antibiotics and form multidrug-resistant bacteria. Solutions are more effective controls on use and disposal of antibiotics which experience suggests are unlikely to be successful. Therefore, it is necessary for treating wastewater from all sources to prevent increasing amounts in the environment and treat drinking water as the direct protection of health. Treatment of wastewater (Samrot *et al.*, 2023) can be tailored to the antibiotics known to be present. Treatment of drinking water has to be a 'catch all' method as there is likely to be a cocktail of substances. Both ozone and activated carbon are highly effective in removing antibiotics from drinking water. Table 2 towards the end of this chapter provides useful information on advantages and disadvantages. Research indicates that these methods can significantly reduce the concentration of pharmaceuticals, including antibiotics, by over 95% (Adams *et al.*, 2002; United States Environmental Protection Agency (EPA)). The city of Zurich, Switzerland held a referendum on the population paying extra for the more expensive treatment. In 2020, a student on the University of Oxford Course on Water Science, Policy and Management for a dissertation project studied whether the population of Oxford would be willing to make a similar voluntary payment. Unfortunately, the project was carried out under Covid lockdown conditions, and it was necessary to make a large number of 'cold calls'. The few people who responded indicated that they would support such a payment. However, it couldn't be considered to be a representative sample.

FOREVER CHEMICALS

There are thousands of man-made forever chemicals (PFAS) in the environment. They do not breakdown in the environment and have become a risk to health (Freedman, 2025). Common ones are firefighting foams and non-stick cookware. They can be removed from drinking water using ion exchange, activated carbon and reverse osmosis. The emphasis is on monitoring, banned production of new substances and banning the use of existing ones such as firefighting foams. See EU and UK legislation (Revolve, 2026; UK Government, 2026). This is resulting in the search for non-PFAS alternatives (Firechief Global).

PLASTIC FIBRES IN WATER

Large plastic objects such as sheets and bags are publicised because of the problems they cause marine mammals and fish, but micro-fibres are everywhere. The concern is the potential health risks of plastic microfibres. They are present in the atmosphere, in rainwater and in water generally. A worldwide survey found that they exist in 83% of drinking water (Chandra & Walsh, 2024). They have even been found in the Antarctic, and in blood samples and in foetuses. Up to 60% of fibres can be removed in conventional treatment of coagulation and filtration and up to 80% with activated carbon, the remaining ones being the smallest ones (<20 ∝m). France has legislated (Alder, 2025) for new washing machines to be fitted with filters and other countries may follow suit.

TREATED WASTEWATER FOR POTABLE USE, FOR IRRIGATION AND FOR INDUSTRY

As discussed in chapters 3, 4 and 5 the need for additional water resources is driving the treatment of wastewater for re-use. Wastewater, often untreated has been used for irrigation. Early civilisations in arid areas, including Egypt, used

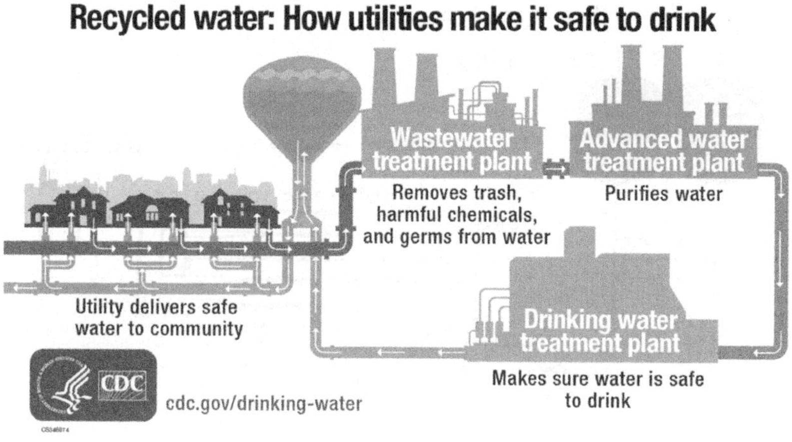

Figure 8.1 See CDC Drinking Water (2024).

untreated wastewater to provide crops with both water and nutrients. Health considerations and standards have driven the need for the water to be treated but other than for expensive crops the use limited by cost as most water for irrigation has been freely available. The need for additional water resources has driven the need for effective treatment to make it safe and meet drinking water quality regulations (CDC Drinking Water, 2024) (Figure 8.1).

Increasingly there are advances in technology (Sharma *et al.*, 2024). For example, for the removal of pharmaceuticals, see Figure 8.2.

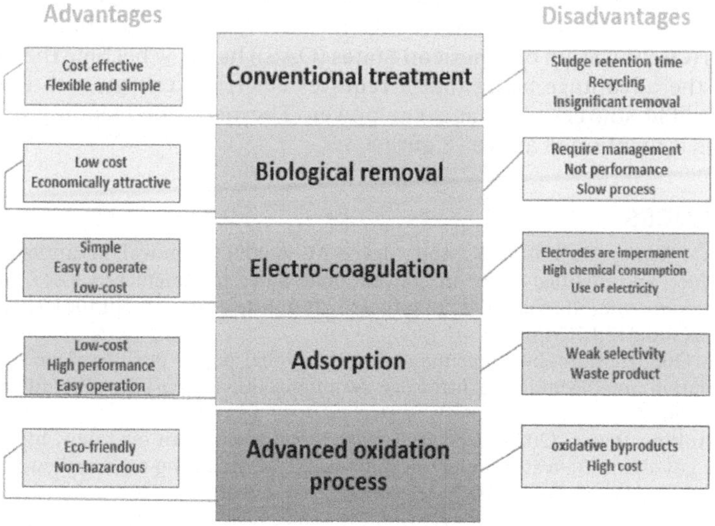

Figure 8.2 Increasing levels of treatment (Source: Sharma *et al.*, 2024).

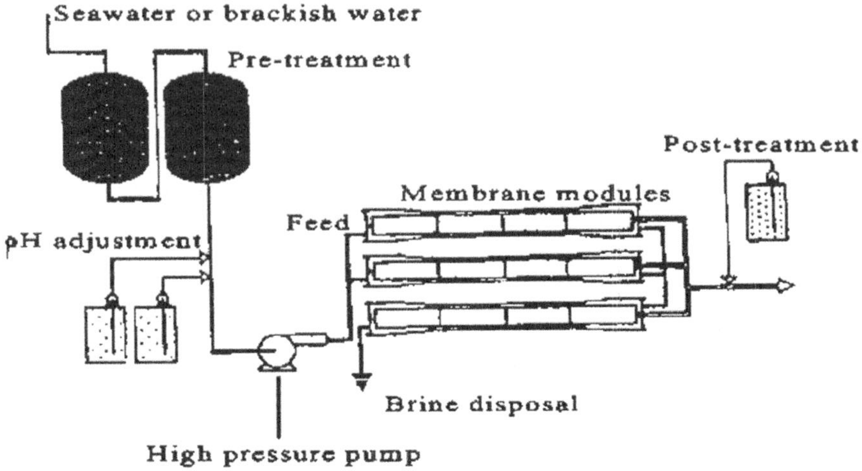

Figure 8.3 Sketch of membrane treatment (Source: Organisation of American States (OAS)).

In Singapore, membrane technologies including reverse osmosis are being used. The disadvantage is the loss of minerals where the treated water needs to be blended with existing water or minerals need to be added. This leads to desalination.

DESALINATION

Traditionally, this involved distillation which required heat and was very petrochemical energy dependent. With advances in membranes, reverse osmosis (Organisation of American States (OAS)) has now become the preferred option, the advantage being that it requires electrical energy with the option of renewable sources such as solar power. The need for blending or adding minerals is mentioned above (Figure 8.3).

REFERENCES

Adams C., Wang Y., Loftin K. and Meyer M. (2002). Removal of antibiotics from surface and distilled water in conventional water treatment processes. *Journal of Environmental Engineering*, **128**(3), 253–260, https://doi.org/10.1061/(ASCE)0733-9372(2002)128:3(253)

Alder M. (2025). Microfiber filter mandates: the global policy push to clean up laundry pollution, intelligent living, https://www.intelligentliving.co/microfiber-filter-laundry-pollution/

CDC Drinking Water. (2024). Recycled water for drinking: an overview, https://www.cdc.gov/drinking-water/about/recycled-water-for-drinking-an-overview.html

Chandra S. and Walsh K. B. (2024). Microplastics in water: occurrence, fate and removal. *Journal of Contaminant Hydrology*, **264**, 104360, https://doi.org/10.1016/j.jconhyd.2024.104360

Firechief Global. Firechief Global, What Are the Alternatives to PFAS Fire Fighting Foam? https://firechiefglobal.com/what-are-the-alternatives-to-pfas-foam/
Freedman A. (2025). Forever chemicals are in your drinking water: Here's how worried to be – and what to do about it. Fortune.com, https://fortune.com/well/2025/06/26/forever-chemicals-pfas-drinking-water-safety-risks/
Organisation of American States (OAS). Desalination by reverse osmosis, https://www.oas.org/dsd/publications/unit/oea59e/ch20.htm
Puertos E. (2025). Water treatment: a guide to processes, importance, and technologies, sustainable business toolkit, https://www.sustainablebusinesstoolkit.com/water-treatment/
Revolve. (2026). EU rolls out mandatory PFAS monitoring to safeguard drinking water, https://revolve.media/insights/eu-rolls-out-mandatory-pfas-monitoring-to-safeguard-drinking-water
Samrot A. V., Wilson S., Preeth R. S. S., Prakash P., Sathiyasree M., Saigeetha S., Shobana N., Pachiyappan S. and Rajesh V. V. (2023). Sources of antibiotic contamination in wastewater and approaches to their removal – an overview. *Sustainability*, **15**(16), 12639, https://doi.org/10.3390/su151612639
Sharma M., Kumar S., Chandrashekar R., Nijhawan G., Muralikrishna C., Habelalmateen M., Goyal R. and Patnaik B. (2024). Modern Approaches in Water Treatment: Emerging Technologies and Future Directions. E3S Web of Conferences, 596, https://doi.org/10.1051/e3sconf/202459601046
Shook. (2024). Ultimate guide to water treatment: everything you need to know, https://www.newater.com/water-treatment-technologies/
UK Government. (2026). PFAS Plan: building a safer future together, https://www.gov.uk/government/publications/pfas-plan/pfas-plan-building-a-safer-future-together
United States Environmental Protection Agency (EPA). Overview of drinking water treatment technologies, https://www.epa.gov/sdwa/overview-drinking-water-treatment-technologies

IWA PUBLISHING

doi: 10.2166/9781789065527_0079

Chapter 9
Early warning systems – EWS

This chapter is in three parts:

Part 1: The Principles of Early Warning Systems
Part 2: Early Warning Systems as part of flood management
Part 3: Early Warning Systems in drought management

PART 1: THE PRINCIPLES OF EARLY WARNING SYSTEMS

There have always been situations when early warning systems could reduce the impact of events, whether it be an earthquake, a landslide or an intense storm, but with climate change resulting in frequent extreme events, early warning systems have become essential.

The definition of EWS and basic requirements is based on the United Nations Office for Disaster Risk Reduction (United Nations Office for Disaster Risk Reduction): 'An EWS is an integrated system of hazard monitoring, forecasting and prediction, disaster risk assessment, communication and preparedness activities, systems and processes that enable individuals, communities, governments, businesses and others to reduce disaster risks in advance of hazardous events'.

Effective governance and institutional arrangements are given in Figure 9.1.

Risk knowledge

It is important to identify the priority risks which are area specific. Risk assessment can identify areas prone to hazard occurrences and locate the nature of vulnerable groups and the critical infrastructure and assets in exposed locations.

Digital mapping, coupled with local knowledge, can map flood risks, providing for the design of monitoring systems and evacuation strategies. The risk information needs to be shared with those affected by the hazards and those organisations responsible for managing the risks.

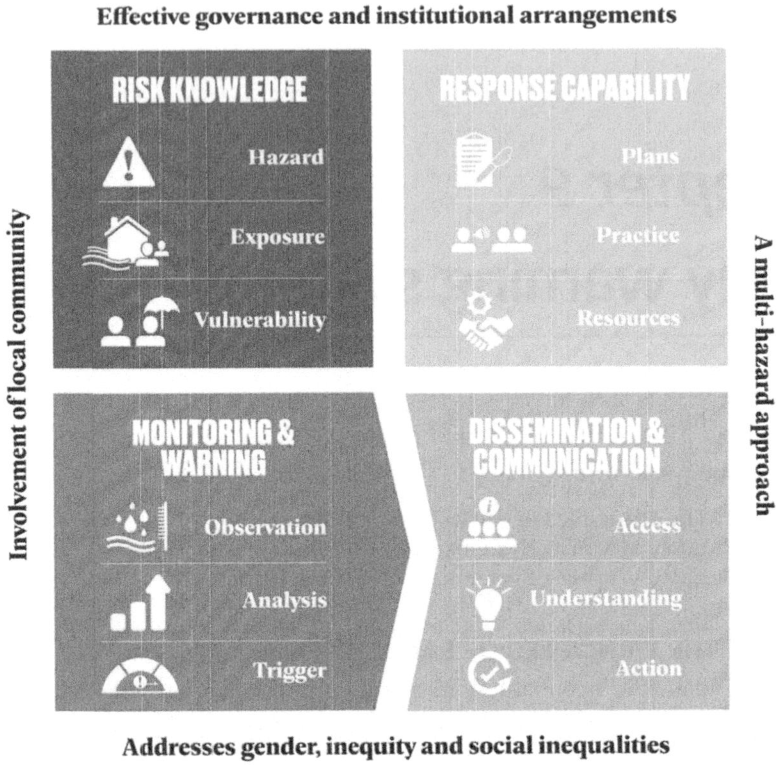

Figure 9.1 Effective governance and institutional arrangements. Advice freely available from a private company, Practical Action (Budimir, n.d.).

Monitoring and warning

Meterological understanding of the conditions which generate a risk of an event, coupled with past experience and event records provide for forecasts of the likelihood of events. The length of reliable lead times is highly area specific with the need for frequent updates leading up to an event. The 'models' need to be able to monitor the movement of the weather system. Hurricanes in the Caribbean are identified several days in advance but their direction and intensity are much less predictable. With experience warning levels and times can be established. Recording and analysis of the outcome of events and warnings issued is important, particularly those in which no warning was issued or in which the warning did not give sufficient time for action.

Dissemination and communication

Governing authorities should agree formal channels on how information and its contents get to the end users. These should achieve immediate receipt with back-ups for absence. Increasingly, because of high ownership, mobile phones are becoming the most common route to the public. The methods of communication and the 'language' used are different for organisations responsible for actions and the public. For example, technical information on weather forecasts is necessary for those making judgements on timing and actions, whereas the public need to know what it means to them regarding actions they need to take. Getting post event feedback from the public is important to be able to improve understanding and maximise the desired responses, particularly in circumstances requiring evacuation.

Effective governance and institutional arrangements

At national level, there should be the overall mandate, legislation and overall funding. Governments should provide the lead and expertise to develop EWS. In most cases this should include continuing involvement to achieve effective systems. In large countries, this could be delegated to area, state or provincial level. Standards, such as warning times, should be set and built into guidance for responsible authorities. Where cross-boundary EWS systems are needed, governments have the responsibility of working together. As needs are constantly changing with climate change, it is important that EWS is kept under continuous review. Public feedback is important, but care has to be taken that such data is not exploited by commercial interests such as insurance companies. Social media has intensified the challenges arising from false information.

Involvement of local communities

Effective EWS (International Federation of Red Cross and Red Crescent Societies (IFRC), 2012) should include the essential elements to consider and prioritise those people at risk. Local governments, NGOs and community organisations should be involved to ensure that an EWS is appropriate for their communities and designed in line with their capabilities. Some communities, such as 'First Nations' with a long passed down generation to generation understanding of weather patterns and signs, are able to provide valuable input to EWS. Community-based EWS systems have been built with the guidance of Practical Action.

Gender and cultural diversity

Women and marginalised groups in some countries are excluded from policies and decision-making. Marginalised people can be overlooked in early warning systems. In this context people without mobile phones are less likely to receive warnings. The community research aspect of an EWS should seek to identify all groups and their communication needs.

PART 2: EARLY WARNING SYSTEMS AS PART OF FLOOD MANAGEMENT

Early warning of flooding is required on a daily or hourly timeframe. In addition, there needs to be long-term planning to reduce rate of runoff and on flood defences. Loss of wetlands, loss of woodland, removal of hedges, farming practice, building in floodplains, lack of provision for rainwater harvesting, failure to build soakaways (swales) alongside roads and from buildings all contribute to greater flooding. Analysing flood events, on where the rainfall flowed from and to, and how those flows could be reduced, can be expected to identify cost-effective flood alleviation investment. It is possible that the impact of increased intensity of some climate change cannot be eliminated, but there are some priority measures which can be identified to slow down the run-off, provide greater reaction time and reduce loss of life and damage. There is a valuable reference: Painter *et al.* (2025).

EWS in practice

The European Union promotes a robust approach to early warning systems, see Figure 9.2.

Text and arrows indicate how the European Flood Awareness System, a preparedness tool, also contributes to the other different phases of the disaster management cycle.

Directive 2007/60/EC (EU Flood Directive 2007/60/EC), the so-called Floods Directive, deals with the assessment and management of flood risk in the EU Member States. It specifically promotes flood forecasting and early warning systems as preparedness action to be integrated in the flood risk management plans. The European Flood Awareness Systems (EFAS) has been designed to close the gap of only partially existing probabilistic

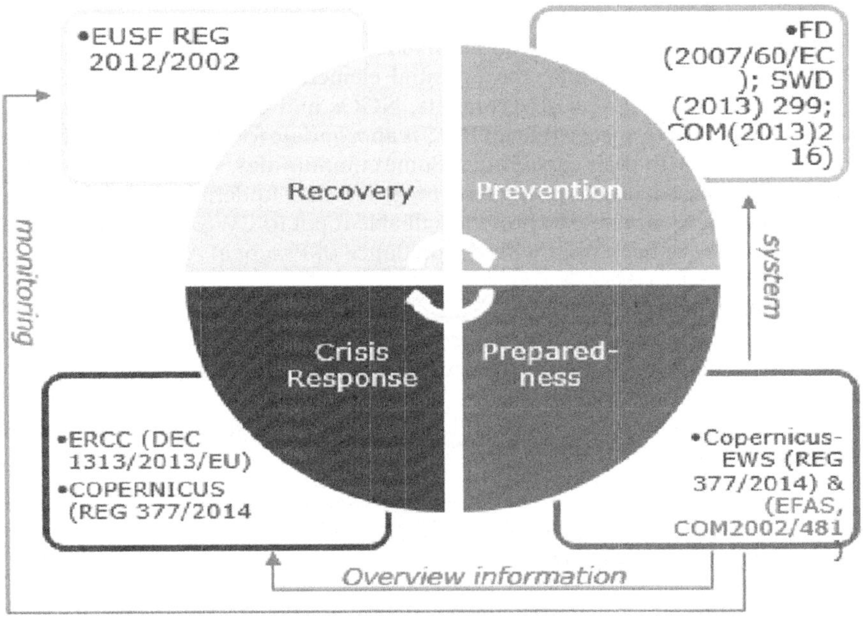

Figure 9.2 EU policies in support of effective flood risk management (EU Flood Directive 2007/60/EC).

forecasting by providing Europe-wide forecasting information with lead times up to 10 days. EFAS use multiple weather forecasts. In total, more than 130 weather forecasts are processed on a daily basis. EFAS does not interfere with national crisis management procedures at any time, so the national services remain the single authoritative voice. EFAS is described in detail in Bartholmes *et al.* (2009).

Flood in Germany in 2021

In 2021 Germany and the Benelux countries suffered the greatest infrastructure damage and fatalities with over 200 people losing their lives in the floods (Zander *et al.*, 2023). Climate change, which made the heavy summer rainfall between 3% and 19% stronger, and 1.2 to nine times more likely, only partly explains why the floods had such a disastrous impact. At least as important was the low level of risk awareness and preparedness and a lack of timely warnings even though there had been an accurate meteorological forecast by the German Weather Service several days in advance of the storm. Communication at that time relied on three weather warning apps. It was estimated that 30–40% of the residents in areas affected by the floods were not reached by severe weather or flood warnings.

Flood in the region of Central Europe 2024

In mid-September 2024, a very large region in Central Europe, including Poland, Czechia, Austria, Romania, Hungary, Germany and Slovakia experienced very heavy rainfall, breaking local and national rainfall records over the period of four days. Almost 2 million people were directly affected by the flooding (Helmholtz Centre for Geosciences (GFZ), 2024) caused by the extreme rainfall with the most severe impacts in urban areas on the Polish-Czech border region and Austria. Emergency management systems across Europe had been reinforced after severe flooding over the last decades. Generally, the systems worked well, despite the higher intensity and larger scale, with the number of fatalities being lower than in earlier floods with 24 estimated at the time of writing, compared with 200 in 2021. It can be assumed that lessons had been learnt from the 2021 event.

Study of three emergency events in Oman (Al-Zaabi & Al-Zadjali, 2022)

There were three different hazards: a tropical cyclone, a tropical depression, and a low-pressure trough. Two sources of data were available: (a) real-time Twitter data that were extracted when events were taking place and (b) formal weather forecasts, advisories, alerts, and warnings by formal agencies. It was found that the three emergencies were not significantly different from each other in terms of reported casualties and property and infrastructure damages, but the warnings against the depression and the trough were delayed and brief compared with the cyclones' adequate warning. The failures were associated with how weather systems are classified in the case study, primarily based on wind intensity. The study recommends the integration of rainfall, storm surge, and flood data when categorising weather hazards.

Practical action experience

The contribution of the consultancy company, Practical Action, was described in Part 1. A valuable example is one on Nepal (Practical Action, 2016). Flood-forecasting methodology was developed for use by the Nepal Department of Hydrology and Meteorology. It used time-series models to generate probabilistic forecasts. The approach is for situations of limited data availability but is different from the impact-orientated approach used in Uganda described below. In the Nepal Karnali River Basin, the forecast increased lead times from 2–3 to 7–8 hours. It stresses the challenge of communicating probabilistic forecasts to communities. The case study paper considers the applicability of the approach to other countries.

Uganda

A study of Uganda by the University of Reading concluded that data was not available to produce an effective warning system, so that a different approach was required. The University considered an impact-based forecasting and warning system. Impact-based forecasting considers the vulnerability of people and property and warns on what the weather will do and what the impacts might be (Mitheu, 2022). Flood monitoring is based on meteorological forecasts conducted by the Uganda National Meteorological Authority, or situation analysis at river-gauging systems. The study found that there was difficulty in identifying the most vulnerable groups of people, but that an impact-oriented approach does allow impact-based early warning systems to be developed through stakeholder consultations. However, validation of an impact-based early warning system using impact data should be done with caution since the quality of these data varies greatly across locations.

Water bankruptcy

Flooding is a lost water resource. Catchment management investment is critical to capture as much rainfall as possible. See chapter 6.

PART 3: EARLY WARNING SYSTEMS IN DROUGHT MANAGEMENT

The principles of early warning systems are given in Part 1 of this chapter. This part is devoted to early warning systems in drought management. Early warning of droughts is largely measured in years but with more urgent updates as the period without rainfall and need for demand reduction measures intensify.

The situation regarding drought management has developed rapidly over the past few years with initial emphasis being on the impact of climate change. Although there had been extreme water shortage situations in arid areas, the two-year drought leading to the 2018 crisis in Cape Town was a wake-up call for all urban areas to plan ahead. Although it had been predicted that a major city would run-out of water, Cape Town had always been regarded as having abundant water supplies. The steps taken from late 2017 until March 2018 are described in *The Last Drop* (Smedley, 2023). They demonstrate the consumption reduction measures necessary to provide just life-sustaining amounts of water. The last step to have been taken in April 2018 would have been to instal standpipes and portable toilets, but this was not necessary as it then rained. In 2018 the Cape Town Drought Response Learning Initiative was launched. A key person in that Initiative was Professor Mark New, Pro-Vice Chancellor for Climate Change, University of Cape Town, and Director of the African Climate and Development Initiative (ACDI). Earlier, he was Professor of Climate Science at the School of Geography, University of Oxford, and the first Academic Director of the University of Oxford MSc Course on Water, Science, Policy and Management. The thinking at that time was that the water crisis could be addressed through effective planning and management. This changed in January 2026, with the publication of the UN Report titled 'Global Water Bankruptcy: Living beyond our means in the post-crisis era' (UNU-INWEH Report: Madani, 2026). This report is discussed in chapter 1. Although this adds to the vital importance of water demand management, it also increases the importance of early warning systems in predicting and sharing limited water resources.

The Water Management Institute, IWMI (Amarnath, n.d.) has been active in drought management for 20 years – see current involvement on the following map (Figure 9.3). IWMI refers to Three Pillars of Drought Management: (1) Monitoring and Early Warning, (2) Risk and Impact Assessment and (3) Risk Mitigation, Preparedness and Response. IWMI is promoting this approach, working with partners around the world under an initiative called 'Drought Action Catalyst' (Institute of Water Management (IWMI)(n.d.)).

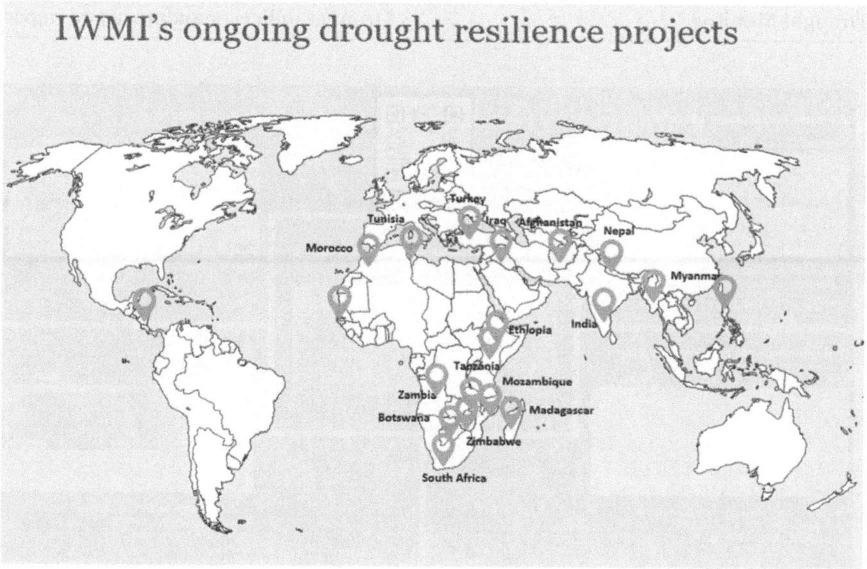

Figure 9.3 Map showing IWMI's worldwide projects on drought management (Amarnath, n.d.).

Giriraj Amarnath, IWMI, presented the IWMI experience to a 2023/4 session of the University of Oxford MSc Course on Water Science, Policy and Management. This provided valuable case study information.

The IWMI approach

This uses the development of an early warning system to identify priorities for drought resilience and early action steps, as illustrated in the following figure and table. Drought early warning and monitoring are crucial components of drought preparedness and mitigation plans. It is necessary to track, assess and deliver relevant information concerning climatic, hydrologic and water supply conditions and trends, and effectively disseminate such information to communities and emergency responders (systematic people-centred early warning systems). Existing institutional arrangements should be strengthened to enhance the resilience of people and communities (Table 9.1).

Table 9.1 Components and outputs taken from Figure 9.4.

Components and Outputs

Component	Output
Weather and Climate Model	(Sub)Seasonal short-term forecasts
Dynamic Drought Contingency Plan	Drought phase indices to activate
Drought Decision Support Tool	Drought alert and active phase
Drought Prediction	Drought prediction using VIC model (a hydrologic land-surface model)
Drought Monitor	Drought indices-mandatory and impact

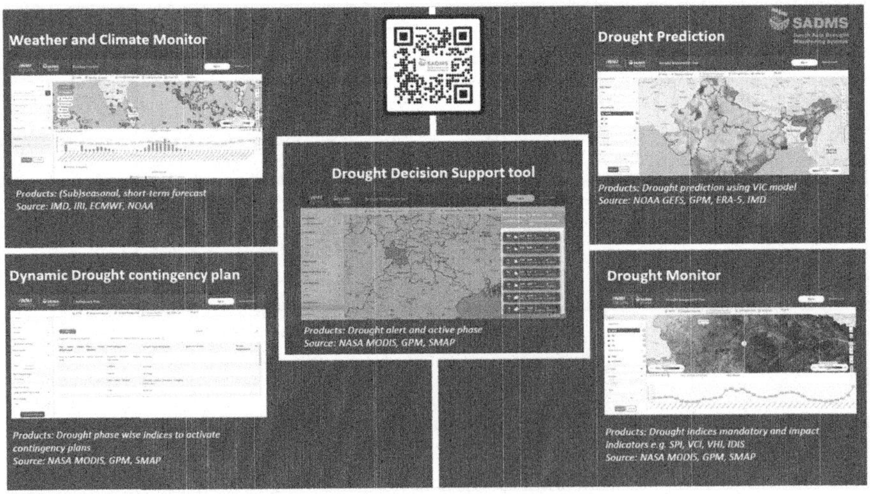

Figure 9.4 IWMI drought action components and outputs (Institute of Water Management (IWMI)(n.d.).

Case studies

(1) IWMI: There are a number of case studies on the IWMI website (Amarnath, n.d.)
(2) Tim Smedley's Book *The Last Drop*. This includes a number of valuable case studies (Smedley, 2023)
(3) State of Odisha, India
The approach RegCDI (Satapathy *et al.*, 2024) is used to identify drought hotspots based on drought severity, duration, and propensity over the study area. A reforecast evaluation of RegCDI (up to three months ahead) showed that the indicators based on soil moisture deficit and crop water stress could predict drought conditions up to two months ahead with no less than 80% accuracy.
(4) Australia
This study presents an operational drought measurement and forecasting system: the Australian Agricultural Drought Indicators (AADI) (Hughes *et al.*, 2025). This system measures agricultural drought impacts via an integrated bio-physical and economic modelling system. AADI (https://www.agriculture.gov.au/abares/researchtopics/climate/) combines existing models including APSIM to simulate crop yields and AussieGRASS to simulate pasture growth, with these outputs then linked to the *farmpredict* model to simulate farm business profits. These models form an operational system, which translates gridded climate observations and forecasts into outcome-based indicators, including winter and summer crop yields, pasture growth, and farm profits.

Transition arrangements

Begin the process of developing or improving early warning systems consulting Practical Action Company as necessary. The development of programmes to reduce the demand for water by communities, irrigation and industry and begin to implement the programmes in drought priority areas. Begin the process of implementation of Drought Action Catalyst consulting IWMI to advise.

REFERENCES

Al-Zaabi S. and Al-Zadjali S. (2022). Qualitative analysis of early warning: a case study from Oman. *International Journal of Disaster Risk Reduction*, **68**, 102731, https://doi.org/10.1016/j.ijdrr.2021.102.731

Amarnath G. (n.d.). Early Warning Systems, Institute of Water Management (IWMI), https://www.iwmi.org/topics/early-warning-systems/

Bartholmes J. C., Thielen J., Ramos M. H. and Gentilini S. (2009). The European flood alert system EFAS – part 2: statistical skill assessment of probabilistic and deterministic operational forecasts. *Hydrology and Earth System Sciences*, **13**(2), 141–153, https://doi.org/10.5194/hess-13-141-2009

Budimir M. (n.d.). Guide to Early Warning Systems, Practical Action, https://practicalaction.org/learning/climate-change/early-warning-systems-saving-lives-and-building-disaster-resilience/

EU Flood Directive 2007/60/EC. EU Flood Directive 2007/60/EC, https://environment.ec.europa.eu/topics/water/floods_en and https://eur-lex.europa.eu/eli/dir/2007/60/oj/eng

Helmholtz Centre for Geosciences (GFZ). (2024). FAQ on the floods in Central Europe in September 2024, https://www.gfz.de/en/press/news/details/faq-zum-hochwasser-in-mitteleuropa-im-september-2024

Hughes N., Gaydon D., Gupta M., Schepen A., Tan P., Brent G., Turner A., Bellew S., Soh W. Y., Sharman C., Taylor P., Carter J., Bruget D., Hochman Z., Searle R., Song Y., Mitchell P., Beletse Y., Holzworth D., Guillory L., Brodie C., McComb J. and Singh R. (2025). Monitoring agricultural and economic drought: the Australian agricultural drought indicators (AADI). *Natural Hazards and Earth System Sciences*, **25**(9), 3461–3482, https://doi.org/10.5194/nhess-25-3461-2025

Institute of Water Management (IWMI). (n.d.). Drought Action Catalyst, https://www.iwmi.org/projects/droughtaction/

International Federation of Red Cross and Red Crescent Societies (IFRC). (2012). Community Early Warning Systems: Guiding Principles. Geneva, https://www.ifrc.org/sites/default/files/CEWS-Guiding-Principles-EN.pdf

Mitheu F. K. (2022). A model for impact-based flood early warning and anticipatory actions in Uganda. Unpublished PhD thesis, University of Reading, UK, https://centaur.reading.ac.uk/112918/1/MITHEU_Thesis.pdf

Painter P., Semmens K., Maxfield K., Cattoën C. and Carr R. H. (2025). Designing effective flood early warning systems: a review of barriers, best practices, and key characteristics. *Journal of Flood Risk Management*, **18**(4), e70145, https://doi.org/10.1111/jfr3.70145

Practical Action. (2016). Flood early warning system in practice: experiences of Nepal, https://infohub.practicalaction.org/server/api/core/bitstreams/cfa4ed0a-c685-447b-9ad7-1535a7ce4331/content

Satapathy T., Dietrich J. and Ramadas M. (2024). Agricultural drought monitoring and early warning at the regional scale using a remote sensing-based combined index. *Environmental Monitoring and Assessment*, **196**(11), 1132, https://doi.org/10.1007/s10661-024-13265-y

Smedley T. (2023). *The Last Drop: Solving the World's Water Crisis,* Picador. ISBN 9781529058147.

United Nations Office for Disaster Risk Reduction (UNDRR), https://www.undrr.org/our-work

UNU-INWEH Report: Madani K. (2026). Global Water Bankruptcy: Living Beyond Our Hydrological Means in the Post-Crisis Era. United Nations University Institute for Water, Environment and Health (UNU-INWEH), Richmond Hill, Ontario, Canada, https://doi.org/10.53328/INR26KAM001

Zander K. K., Nguyen D., Mirbabaie M. and Garnett S. T. (2023). Aware but not prepared: understanding situational awareness during the century flood in Germany in 2021. *International Journal of Disaster Risk Reduction*, **96**, 103936, https://doi.org/10.1016/j.ijdrr.2023.103936

doi: 10.2166/9781789065527_0089

Chapter 10

Financing aspects, timetables and priorities

This chapter considers the required investment framework, what is required to be funded identified in the foregoing chapters. It distinguishes between public goods, regulated water service delivery investment, household/customer-funded services and commercial. On timescales it considers immediate requirements, the medium-term needs and the long-term infrastructure and institutional investment.

BACKGROUND

A reminder of the situation in the era of water bankruptcy

Beginning with the Water Decade in 1980, there have been a number of attempts to stimulate the development of water services by setting goals on what needs to be achieved. Although the intentions were good, very little progress has been made with over 2 billion people worldwide without access to safely managed drinking water in 2025. There are many reasons for this, such as lack of political commitment, lack of finance and a lack of attention to how the necessary improvements might be made. The report of the UN University on Water in Canada (Madani, Global Water Bankruptcy. UNU-INWEH Report, 2026) was a wake-up call by announcing the post-crisis era of water bankruptcy. This did not relate to climate change but to the growth of the world's population now exceeding the rainfall to support it. The chapters in this book address the challenges to be faced in this era. It is recognised that there will be different situations across the world.

Water investment must be made in spite of financial and conflict turmoil

In a dangerous world few would argue the case for defence funding but how does that compare with that required for water? It is difficult to compare data from different sources, but it is possible to make broad comparisons. The global total defence spending in 2025 was 2.9 trillion US dollars (MilitarySpend.org, n.d.). Much of this had been made available at the detriment of AID budgets. The WaterAid led project (WaterAid, 2022) gave various estimates suggesting that water investment needs over time are in the range of 7%–14% of those 2.9 trillion dollars. It is important to consider the differences, defence spending is largely for now with a significant proportion on consumables such as ammunition, drones and rockets, and less on infrastructure such as air raid shelters, whereas much of water spending is for long-life infrastructure. The immediate investment requirements in water are relatively low with greater amounts required for long-term infrastructure.

INVESTMENT FRAMEWORK

Factors influencing decision making

(1) Costing but not valuing water: The fundamental difference is that cost is what you spend on goods or services whereas value is subjective on the perceived benefit. The cost say of a service can be quite low – an electrician replacing a faulty cable compared with the value of preventing a house fire. In water, farmers often get water free but with value potentially as high as business survival. I give two famous quotes, Warren Buffet: 'Price is what you pay, value is what you get'. Quite differently, Oscar Wilde describes a cynic as someone 'who knows the price of everything but the value of nothing'. Water is underpriced because it is not valued.

(2) Tariff policies, subsidies, borrowing and life-cycle asset replacement: Tariff policy should be established to achieve service sustainability. In practice because unexplained tariff decisions are unpopular, they are set to minimise the public reaction. It is interesting that in Switzerland (e.g. in Zurich) tariffs have to be agreed through referenda. As the reasons and benefits are understood, sustainable cost recovery is achieved.

General subsidies result in water service delivery organisations becoming inefficient, resulting in a poor service and inadequate attention to system maintenance and renewal. Subsidies targeted at the poor can be valuable, but it is important to understand affordability. Because the poor tend to receive an inadequate service, due to lack of investment in developing distribution networks, they end up paying perhaps 15 or more times than the tariff rate due to having to buy water from vendors.

A lack of investment over time results in the water infrastructure being unable to meet rudimentary service standards. An old adage is 'a stitch in time saves nine'. Something that the governments seem to ignore whether maintaining roads before potholes occur faster than they can be repaired, or investment in repairing leaks in a distribution system before supplies become intermittent. Investment to achieve 24/7 supplies is essential (see chapter 3) and should be a priority in the longer-term infrastructure investment programme. See Timetable and Priorities at the end of this chapter.

(3) How to avoid deferred maintenance, debt accumulation and politically unsustainable water service pricing: The historical view of water needs to be overcome through education. The world would not exist without

water and a climate that makes that possible. Long before we had commerce people lived near a water source and took it freely. Later, with growth in the population and people congregating in 'cities' it became necessary for water to be collected and distributed which involved water infrastructure and costs. In many cases there was no charge for water and the attitude prevailed that water should be free. A good example of effective public communication; I am reminded of the story in Swaziland which I give in chapter 2 and I reproduce it here. The Minister responsible for water in Swaziland put this argument in perspective at a public meeting, when he was reported as saying words to the effect that 'certainly water is free, providing you wish to take your bucket up the mountain to the source to collect your water, but if you wish me to make provision for the collection, treatment and delivery of safe drinking water to your home, then I have to charge you for the service'. Water is a public good and against this background governments have tended to underprice water services and haven't marketed it based on their value. Related to this is that water infrastructure has a long life, much longer than the tenure of most administrations, such that its replacement has been neglected leaving a financial legacy coupled with the need for additional new systems. An education programme should begin as early as possible covering politicians, local government officials, water service payers, schools and the media. The water bankruptcy message can be prominent in this endeavour, but the task is the most difficult of all the challenges as it involves people not things. Enlisting the assistance of influential public commentators would be valuable. Chapter 7 discusses communication and public participation.

Funding

	What Needs to be Funded	Funding Options
1	Data collection surveys	Public, Regulated utility investment
2	Catchment/nature-based solutions	Public, Household-funded service
3	Crop development	Public, Commerce
4	Urban network Infrastructure	Regulated utility investment, Public
5	Safely managed drinking water	Regulated utility investment, Public
6	Wastewater and drainage	Regulated utility investment
7	Capacity building and training	Public, Commercial

Discussing each in turn:

(1) **Data Collection Surveys:** In many parts of the world there is a lack of information on parts of the infrastructure and its condition, such as water supply distribution systems. Without obtaining the information it will not be possible to determine the development needs and costs. Funding collection contracts can be undertaken by the utility supplier,

or may rest with governments. The data collection contract can be part of a long-term concession management contract.

(2) **Catchment/nature-based solutions:** Modifying river flow can require substantial earth moving which most likely would be government funded, but tree-planting (The Conservation Volunteers, 2025) is often carried out by 'volunteers for general benefit, but also may increase the value of property, reduce flooding-related insurance and have mental-health benefits.

(3) **Crop development:** As discussed in chapter 4, there are large financial benefits in switching to crops more suitable to the local climate with significant reduction in water requirement and increased income for farmers. The necessary research is likely to be carried out by academics but there are significant commercial opportunities to be invested in by farmers and their customers. Also, there are potential benefits in relation to virtual water if for example, a coffee bean were to be developed able to grow in temperate climates. Perhaps there could be interest in investment by supermarkets.

(4) **Urban network infrastructure:** In many parts of the world the infrastructure has not kept pace with the expanding cities with population growth and rapid urbanisation. The responsibility will ultimately rest with the utility but either the investment cost will be too high to be recovered in service charges requiring government subsidies or parts of the infrastructure itself has to be funded by government.

(5) **Safely managed drinking water:** This requirement has been waiting for concerted action since the Water Decade of 1980–2000. In the era of water bankruptcy it is now a critical requirement. It must be actioned. There are options, the main one being a conventional water treatment and distribution system. An alternative is the distribution of treated water to community storage tanks or to kiosks from where the water can be collected. This should be an enforced regulatory requirement.

(6) **Wastewater and drainage:** Around the world habitations were established nearby to water sources, often rivers. Overtime with city developments, those original dwellings have been overwhelmed by flooding recently exacerbated by climate change with more severe and more frequent storms. The necessary investment in wastewater treatment and surface flow has not kept pace. This is a clear responsibility of wastewater utilities and action should be regulated and enforced.

(7) **Capacity building and training:** This book has identified the need for up-to-date knowledge and skills and research and development, for example to identify drought-resistant crops. Governments, operators, consultants and other commercial organisations will require trained individuals with the ability to carry things forward, so the early period should include investments in training facilities. Governments should ensure that training facilities are in place. Some countries have water-based training institutes. Opportunities for less-developed countries should be offered such as internships with those who have world

experience. There are opportunities for consultants to work worldwide to work with countries who need guidance with the prospect of contracts. NGOs, especially those water-based with worldwide experience such as WaterAid have a major role to play.

Blended finance

The definition of blended finance by The Organisation for Economic Co-operation and Development is 'The strategic use of development finance for the mobilisation of additional finance towards sustainable developments in developing countries'. In practice, it has not been established widely. Successful blended finance projects yield impressive benefits relating to around a million people. Two examples from a case study (Tkachenko & Mansour, 2021): a project in Bangladesh yielded sanitation loans to 776 000 people; a pipeline development in Cambodia benefitted 1.1 million people. The WaterAid-led project on 'Mobilising Capital for Water' (WaterAid, 2022) also has some valuable case studies. It provides a valuable list of possible funding models for water investment sub-sectors. Catchment sizes around the world vary to a rural community size to whole river systems. In chapter 6 I suggest that long-term managing rainfall catchment may require local responses based on communities. Blended finance should be considered where it is appropriate for water projects, namely with bankable revenue streams, strong regulation and well-defined service contracts.

TIMETABLES AND PRIORITIES

As has been proposed throughout this book a transition period will be necessary, beginning with awareness and planning to a new norm of continuous investment. There is a need to consider immediate requirements, the medium-term needs and the long-term infrastructure and institutional investment.

Immediate requirements

The transition in a world in turmoil to a world with governments recognising the absolute need to make a commitment to finance water is a major step. Once that commitment is made there would be an incentive to take some immediate steps and to plan for the medium term and the long term.

The immediate steps for all governments are to

(1) Set up a communication system which reaches all sectors of populations to explain the need for action and why the decision has been taken, what initial actions are being taken, how they will be consulted and how they will be kept informed.
(2) Set up a planning / monitoring group (action orientated not talking) within government involving all departments with a clear lead unit, and involving advisers from all sectors, local government, water utilities, financial bodies, consultants, industry and academia.
(3) Place contracts to identify the system data requirements and to prepare data collection contracts.

(4) Outline the reform needs to be carried out in the medium-term programme.
(5) Develop a vison for the future to give an aim for the long-term.

Medium-term and long-term

To incorporate the items listed in the table giving the list of requirements. The planning taking place in the immediate requirements would be likely to trigger some earlier action on some items.

REFERENCES

Madani K., Global Water Bankruptcy. UNU-INWEH Report. (2026). Global Water Bankruptcy: Living Beyond Our Hydrological Means in the Post-Crisis Era. United Nations University Institute for Water, Environment and Health (UNU-INWEH), Richmond Hill, Ontario, Canada, https://doi.org/10.53328/INR26KAM001

MilitarySpend.org. (n.d.). World Military Clock – Real-Time Global Defense Spending, https://militaryspend.org/world-military-clock

The Conservation Volunteers. (2025). I Dig Trees programme: 9 Surprising Benefits of Planting Trees in Your Community (Backed by Science), https://www.tcv.org.uk/media_hub/benefits-of-planting-trees-community/

Tkachenko O. and Mansour G. (2021). Case Study Assessment: Blended Finance in Water, Sanitation and Hygiene (WASH) – Lessons for Development Partners. International Water Centre, Griffith University, Lean Finance and Aguaconsult, Australia, https://watercentre.org/wp-content/uploads/2024/04/Blended-Finance-for-WASH-Lessons-for-Development-Partners-Final-Report-Public.pdf

WaterAid. (2022). Mobilising Capital for Water: blended finance solutions to scale investment in emerging markets, https://www.wateraid.org/washmatters/resources/mobilising-capital-water-finance-solutions-scale-investment-markets

doi: 10.2166/9781789065527_0095

Chapter 11

England and Wales

Whatever the system, consistent government direction is essential. The failure of regulation of private water companies in England is an important lesson to be recorded. On establishing the private companies in 1990, with the drive of the private sector ethic controlled by independent regulation, it looked to be a progressive development. Why it failed and where we are in 2026 is discussed below.

In October 2024 the UK Government set up a review of regulation by an independent Water Commission chaired by Sir Jon Cunliffe. The Cunliffe report (Independent Water Commission: review of the water sector (Cunliffe Report), 2025) was published in July 2025. It was very comprehensive, 464 pages long with 88 recommendations. Based on my experience as head of the Drinking Water Inspectorate during the period 1993–2003, the reasons for the failure of the regulatory system are the following:

- In the 1989 Act (Water Act 1989) which set up privatisation, there was no provision for river quality goals or enforcement, so there were no targets set for investment in river water quality. Later, the opportunity presented by the EU Water Framework Directive (EU Water Framework Directive) in the year 2000 to add the provision to achieve good quality river water was not taken.
- The National Rivers Authority (NRA) which was established as part or privatisation for regulation of rivers had powers only to prosecute pollution incidents.
- Later, Government's attention was focused on competition, as had been introduced for telecom and energy services. The aim was to give customers a choice of supplier based on quality of service and level of charges.
- The Economic Regulator, Ofwat, responded by staffing regulators from the energy industry with competition experience. These staff had no knowledge of water services.

- On setting up privatisation there were many shareholders including water customers, providing some public involvement in ownership. These were bought out on the argument by the water companies that they were administratively expensive. The impact was the loss of 'internal' scrutiny.
- The assumption (perhaps hope) at that time of privatisation was that new capital would be obtained through additional equity. In practice, much of the investment money needed was obtained through borrowing with the water companies able to increase their profits by obtaining interest rates lower than those set by Ofwat.
- This was the beginning of the water companies accumulating debts. Later, Ofwat allowed water companies to be taken over by banks, whose only interest was in lending money. In 2017 a report by a BBC news programme (BBC News, 2017) which had investigated the pollution of the River Thames, reported that the Australian Bank, Macquarie, which had owned Thames Water from 2006 to 2017, had left the company with an additional debt burden of £2 billion. This was disputed by the bank at the time. The investors, who bought Thames Water from Macquarie, did not buy out the debt. This debt remained ahead of the Cunliffe Review.
- When the National Rivers Authority became part of the EA dedicated water oversight was lost. Water had become just one part of environmental regulation.
- When Government, as part of government austerity measures, slashed EA staff numbers there were insufficient staff to monitor river quality.
- River users, some of whom were keen river swimmers, began to carry out their own tests of river quality. Thames Water tried to hide this by issuing masses of data supposedly related to sewage discharge quality. This was uncovered by a former police detective superintendent, and founder of 'Windrush Against Sewage Pollution', Ashley Smith who discovered that Thames Water had not been monitoring sewage effluent discharges nor inspecting sewage treatment works. Ashley and fellow campaigner Peter Hammond challenged the water company at every opportunity and the information uncovered was presented in a Channel 4 drama in February 2026 (Channel 4 Drama, 2026).
- Public pressure resulted in the setting up of the Cunliffe regulatory enquiry.
- The result of the above government and regulator failures was neglect of investment in wastewater treatment and storm overflows. Aside from the water service provider failure, the effect is seen in greater flooding. The country is left with both environmental problems and a large investment backlog.
- In comparison, the Drinking Water Inspectorate with enforcement powers was able to require investment to be made to achieve the standards of the EU Drinking Water Directive.

Some key points in the Cunliffe Report:

- There should be a 25-year national strategy.
- There should be just one integrated regulator covering economics, environmental and drinking water quality regulation. Within that regulator, the approach to drinking water regulation should be preserved.

- There should be river quality standards based on the EU Water Framework Directive and the recently updated Wastewater Treatment Directive.
- The required enforcement of river quality was implied but not specified.
- Renationalisation was ruled out due to cost.
- The future form of the nine private water companies in England was left to a company-by-company decision if and when a company was put into administration.
- An Implementation Advisory Group was recommended. This was seen as essential as there are many decisions to be made and details to be worked through.
- One key aspect is to achieve joined up water and city planning to achieve integrated investment through government on the catchment and through water companies on wastewater treatment and in addressing stormwater overflows.

ENGLAND ONLY

Welsh Water (Dŵr Cymru) was part of the original privatisation but, to avoid being bought by an American company, it changed its status to a 'mutual' which has trustees instead of shareholders. Although it was subject to the same regulation vagaries of English companies, it has emerged in better condition. Because of Welsh Water's non-private status, the remainder of this chapter relates to English companies only.

Could English water privatisation have been successful with effective regulation? The fundamental flaw is that Public Limited Companies (PLCs) are unsuitable for public services because they can't be allowed to go bust. The following would have been required:

- Government to have established clear river quality goals and not become sidelined by an obsession with competition.
- Ofwat to have been given clear guidelines on water company ownership related to equity-based investment and returns related to customer service and standards delivery.
- River-quality regulation to have been properly funded.
- To have had integrated water and city planning.
- A big 'if' would be if financial penalties on water companies for delivery failures were directed at shareholders and management, for example, in regulating bonuses. A private company may not be willing to accept such regulatory conditions.

Government response to the Cunliffe Report

A white paper was published on 20 January 2026 titled *A New Vision for Water* (UK Government, 2026). The Government hails *A New Vision for Water* as the largest overhaul of the water sector in a generation. The reforms aim to transform the industry into a 'low-reward, low-risk, long-term' model to tackle pollution, fix ageing infrastructure, and restore public trust.

The key aspects of the white paper are summarised below:

(1) **Structural and Regulatory Overhaul**
- Single Regulator: The government proposes abolishing Ofwat and creating a new, integrated regulator that merges the functions of Ofwat, the Drinking Water Inspectorate (DWI), the EA, and Natural England.
- 'Pre-pipe' focus: a shift from reactive pollution management to proactive, 'pre-pipe' solutions, such as sustainable drainage and tackling sewer misuse.
- Supervisory approach: the new regulator will move away from industry-wide benchmarking towards 'company-specific' monitoring, with teams tailored to each water company.
- Chief engineer: a new chief engineer role will be established to provide engineering-based supervision of assets.

(2) **Investment and Infrastructure**
- £104 billion investment: the plan is backed by a projected £104 billion in private investment over five years (2025–2030).
- Long-term planning: the 5-year price review cycle will be supplemented by a 25-year delivery framework to encourage long-term planning over short-term profits.
- Resilience standards: introduction of statutory resilience standards and a focus on upgrading infrastructure to tackle storm overflows and wastewater treatment.

(3) **Environmental and Pollution Measures**
- End to self-monitoring: the white paper commits to ending operator self-monitoring by water companies, replacing it with a new 'open monitoring' approach using automated, real-time data.
- Tackling agricultural pollution: a commitment to increase farm inspections to at least 6000 annually by 2029.
- Nature-based solutions: greater emphasis on using nature-based solutions, such as wetlands, for water management.

(4) **Customer and Ownership Protections**
- Water Ombudsman: a new independent Water Ombudsman will be created with binding powers to resolve consumer complaints.
- Water poverty: a target for water companies to eliminate water poverty by 2030.
- Ownership oversight: the Government will introduce mandatory notification under the National Security and Investment Act for any significant changes to water company ownership, and develop a transparent process for assessing 'not-for-profit' models.

(5) **Next Steps**
- Transition plan and Water Reform Bill: a 2026 transition plan will provide a roadmap for these reforms, followed by a Water Reform Bill to implement the necessary legislation.

Immediate Action: while the new regulator is established, current regulators (Ofwat and EA) are expected to align with these reforms in the interim.

What is not mentioned in the White Paper:

- Although some restoration of wetlands is mentioned, there is no discussion of the need to control rainfall runoff to reduce flooding. For integration of planning the Cunliffe Report recommended the formation of Regional Authorities. These are not mentioned.
- The protection of the Drinking Water Inspectorate's independence of approach to regulating drinking water quality.
- River quality target bases, such as the EU Framework Directive and the Urban Wastewater Treatment Directive.

At the time of preparing this book, the transition plan is awaited.

TRANSITION ARRANGEMENTS

The Government should consider whether the provisions in its White Paper produced to take forward the recommendations in the Cunliffe Report are relevant and sufficient to address the challenges of water bankruptcy. Focusing on controlling run-off and capturing rainfall will require an integrated approach. It will be necessary to decide whether there will be a requirement for a change in the scale of operations of the water companies.

Wastewater legislation

Alongside the activities relating to water regulation and organisation change, wastewater legislation was prepared earlier but not enacted until following the regulatory and water company failures, which lead to the Cunliffe enquiry.

The Government's response to flood events is to prevent developers from connecting any additional stormwater connections, and for sustainable drainage systems (SUDS) to be incorporated. A recent UK Government statement (May 2025) (UK Government, 2025) on capacity of sewers and sewage treatment required water companies to handle both sewage and rainwater. The issue of stormwater overflows is described in chapter 6. The aspects to be addressed are:

- Recreating separate sewers.
- Government to force developers to build only separate systems, including building soakaways for roof and site rainfall.
- Discharge stormwater to swales which are roadside rainwater soakaways described in chapter 6.

What steps are water companies taking?

- Thames Water has built the Thames Tideway Tunnel (Institution of Civil Engineers (ICE), n.d.), operating since February 2025 to collect 34 combined-sewer overflows at a cost of £5 billion. This will greatly reduce river quality problems in London, but at time of writing no actions have been taken to address the severe stormwater overflow problem elsewhere.

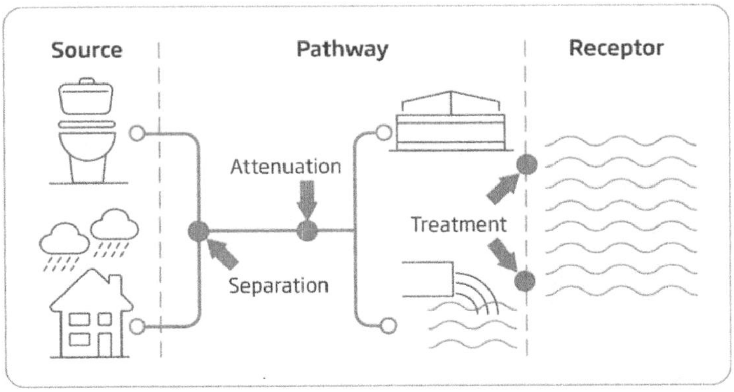

Figure 11.1 Separation and Treatment of Sewage and Rainwater Drainage Systems.

- Wessex Water has taken a holistic approach (Wessex Water, n.d.). In its 2025–2030 storm overflow programme it is:
    - Spending £515 million on 143 overflows to reduce their impact on the environment; this includes 31 nature-based solutions.
    - Increasing capacity of wastewater treatment plants.
    - Constructing storage tanks that will hold the water for later treatment.
    - Where practical, separating sewage and rainwater.
    - Reducing groundwater infiltration.
    - Using nature-based systems, such as wetlands and reedbeds to treat storm overflows (Figure 11.1).

REFERENCES

BBC News. (2017). How Macquarie bank left Thames Water with extra £2bn debt, https://www.bbc.co.uk>news>uk-60040162

Channel 4 Drama. (2026). 'Dirty Business'.

EU Water Framework Directive. EU Water Framework Directive, https://environment.ec.europa.eu>topics>water>water-framework-directive_en

Independent Water Commission: review of the water sector (Cunliffe Report). (2025). https://www.gov.uk/government/publications/independent-water-commission-review-of-the-water-sector

Institution of Civil Engineers (ICE). (n.d.). Thames Tideway, https://www.ice.org.uk/what-is-civil-engineering/infrastructure-projects/thames-tideway

UK Government. (2025). About the guidance and planning for drainage and wastewater management, https://www.gov.uk/government/publications/guidelines-for-statutory-drainage-and-wastewater-management-plans-dwmps/about-the-guidance-and-planning-for-drainage-and-wastewater-management

UK Government. (2026). 'A New Vision for Water' White Paper, https://www.gov.uk>government>publications>a-new-vision-for-water-white-paper

Water Act 1989. The Stationery Office, ISBN 0105415898.

Wessex Water. (n.d.). Storm overflow investment, https://corporate.wessexwater.co.uk/our-purpose/rivers-and-coastal-waters/storm-overflow-investment

doi: 10.2166/9781789065527_0101

Chapter 12

Summary of critical elements

This book aims to address the major challenges faced not just by a worldwide water crisis, but the more severe and urgent matter of water bankruptcy. The book does more than consider 'the what' but also 'the how', covering both well-established practices and suggested approaches to matters which have not been addressed. In this summary I stress just a few critical elements.

(1) At the time of writing the world is embroiled in political turmoil with wars and economic instability. Few would argue against defence spending in such dangerous times, but before this demand on government budgets, much of the available finance required in developing countries was made available through Aid Budgets. The Stockholm International Research Institute estimates that world spending on defence in 2025 was 2.9 trillion US dollars. It is difficult to get a meaningful figure for water to make a comparison with that defence figure. However, it is important to make an assessment as much of the defence monies have been budgeted by those countries cutting their Aid Budgets, of which some would have been applied to water developments. The required water percentage could be between 7% and 14%. Timetables for investment are also important. Transitional arrangements have been discussed in each chapter. The immediate water funding requirement will be relatively low, but with longer term infrastructure investment needed.

Since the recent world turmoil there has been little or no discussion on water management. I was delighted to see an article in the South China Morning Post on 18 May 2026 (author: Ambika Vishwanath) headed 'South Asia must make water management a top priority'. Essentially, the message in that article is whatever the political uncertainties, water must take top priority. With political will, increasing spending on water should be possible, providing that its importance is recognised. It is no longer feasible to 'bury heads in the sand'. What needs funding and funding options are discussed in chapter 10. NGOs and charities often provide stimulus and guidance. A well-established one is WaterAid based in the UK which, working with local organisations, delivers worldwide improved water and sanitation to communities which have the greatest need.

(2) The overriding physical requirement is to increase the capture of rainwater and when feasible store it, and to reduce water consumption in communities, irrigation and by industry. Some chapters, specifically 3, 4, 5 and 6 address these aspects directly. Other chapters cover the necessary enabling aspects such as planning and finance, and associated factors such as water quality which reinforce the overriding hydrological requirement.

(3) With greater variability of rainfall and increasing intensity of storms, early warning systems for floods and droughts need to be developed with ongoing review of systems based on experience. The risks are widespread with nowhere in the world being risk-free. Countries with similar circumstances should put in place procedures to share experience, such as takes place in the European Union.

(4) Public participation is vital, not just as a human right and the need for public buy-in, but also because people can make a major input through volunteer effort especially on activities in natural water catchments. I hope that advocacy individuals and groups will hold governments to account for action programmes. Each individual should consider WHAT CAN AND SHOULD I DO?

Taking early action is imperative. Unless the worldwide water situation is addressed with concerted and continuous action, there will be a water catastrophe with the likelihood of millions of deaths from famine.

doi: 10.2166/9781789065527_0103

Appendix

INTRODUCTION

The appendix provides key aspects of chapters in my book *Institutional Governance and Regulation of Water Services: The Essential Elements, 2nd edition* (2013), which, although valuable in their own right, don't align with the focus of this book.

I see some form of independent regulation as essential for sustainable water services. There are those who believe that integrity in local government is sufficient, and others who believe that "the market" can achieve everything, and they would disagree with my statement. I leave it for them to justify their arguments. This chapter gives my arguments as to why regulation is essential, and why some form of independent regulatory body is necessary in most situations. There is a great deal of scepticism as to whether regulators can be truly independent, and this depends critically on the political climate in a country, and on the will of the people. Thus, the relationship of regulators with public representative bodies is critical. It is important at this stage to stress that "one size doesn't fit all", and the appropriate form of regulatory body depends upon a country's situation or the local circumstances.

Regulation of water services only really became an international issue following the "privatisation" of the Regional Water Authorities in England and Wales in 1989, and the interest this stimulated in the greater use of the private sector elsewhere. There arose a concern over the control of private monopolies and the UK Government used the experience of privatisation of the energy and other sectors to introduce a regulatory regime. A 1997 paper [1] on regulation of privately operated water services said "there is widespread agreement that independent regulators are at the heart of regulatory governance for liberalised sectors", referring to OECD, the World Bank, IMF and Regional Development Banks. Due to the ineffectiveness of some regulatory bodies, more recent papers

[2] are now questioning that earlier statement, and I consider the competing arguments later in the chapter. However, in 1997, although there was enthusiasm for regulation of the private sector, there was less or even no recognition of the need for regulation of the public sector. As will be seen in Chapter 7, there is now a growing recognition of an equal need for regulation of publicly managed water services to achieve both transparency and efficiency. The UN Human Development Report 2006 [3] referred to the need for regulation of publicly managed water services, with one of its recommendations for government action being "regulating water utilities to improve efficiency, enhance equity and ensure accountability to the poor". In this chapter, I make no distinction between regulation of public and private operations.

There are four areas of regulation. Three of these can be described as economic regulation, drinking water quality regulation and environment regulation, all of which are covered in this book. The fourth area, relating to the enforcement of general business and commercial law, such as those associated with accounting practice, employment legislation and health and safety at work, is not covered, as it is not peculiar to water services. This chapter covers the broad canvas of regulation and regulators and Chapters 3 and 4 cover the more detailed aspects of planning, finance and cost recovery, and standards, monitoring and enforcement respectively. We need first to have a clear view of what we mean by regulation.

WHAT IS REGULATION?

Regulation has been around for as long as we have had laws that relate to the protection of citizens. Police are monitors and enforcers. In 1985, Selznick [4] described regulation as "sustained and focussed control exercised by a public agency over activities that are valued by a community". Tenenbamn [5] in 1995 referred, in a power sector context, to "regulation (as) a system that allows a government to formalise and institutionalise its commitments to protect consumers and investors". Baldwin and Cave [6] point out that although regulation is seen generally as a means of control of activities to prevent something undesirable from happening, regulation can also be a means of facilitating something good such as efficient and disciplined use of radio frequencies. This is important, not just in the form of regulation as in the Baldwin and Cave example, but in the way in which a regulator operates. Within "a command and control" regulatory regime there is a balance to be made between the use of the carrot and the use of the stick. Not only does the carrot produce quicker results, but it prevents pressures developing for deregulation. Regulation managed in a "sensible" way can be seen as desirable, rather than a burden, to those being regulated. I will return to this later. Mcintosh [7], in the context of the development and management of urban water supplies and sanitation in developing countries, sees "regulation (by definition) as being about making and enforcing rules". Jamal Saghir [8] warned that "regulation is not a substitute for governance". I agree with that but, at the beginning of Chapter 1, I gave a list of elements which together I see as constituting good governance, and two of these relate to regulation. One is "regulatory bodies that are independent and have the power to implement policies, laws and rules in a transparent way", and the other is "effective monitoring and enforcement of regulations". I see regulation as the process of interpreting and implementing laws, policies and regulations, to achieve what was intended in their formation. This is a pragmatic definition which can be at odds with the views of lawyers. There is another reason why the carrot is better than the stick. The stick can result in litigation in an enforcement process with undesirable outcomes. The use of courts by a water regulator is essential at times, but I see it as a failure, by a regulator, to persuade those it regulates to meet the requirements, when it has to resort to legal sanctions. The "carrot and stick" argument is complicated by the attitude of the media in assessing the reputation of a regulator according to its prosecution "score". This can be compounded by pressure groups who wish to have their "pound of flesh". So the general definitions of regulation show that it can cover a wide range of different aspects. In the water sector, even though there is more quality than economic regulation, the word regulation can be seen as economic regulation, so it is important now to give a specific definition to that.

WHAT IS ECONOMIC REGULATION?

It is important to distinguish between the wider roles of regulators and the aspect which is economic regulation. The World Bank Explanatory Notes [3] discuss "economic regulation versus regulation generally", and conclude with a core definition of economic regulation as "the rules and organisations that set, enforce and change the allowed tariffs and service standards for service providers". The Notes give a valuable illustration of the overlaps with other aspects of regulation, with the source attributed to Castalia, which I reproduce in Figure A.1. The core definition covers only the non-overlap portion labelled monopoly, but in practice economic regulators also have varying responsibilities for some of the overlaps. I discuss the advantages and disadvantages of aggregation of these functions in a later section.

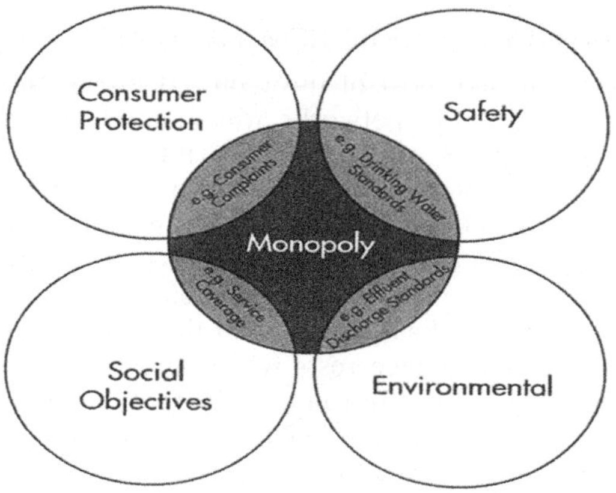

Figure A.1 Overlap between regulatory functions (*Source*: Castalia).

WHY NOT SELF-REGULATION?

Eijlander [9] gives three characteristics of self-regulation: it concerns the regulation and ordering of behaviour in a certain group of society; it consists of rules which have been laid down for and work within the group; and these rules can be enforced against the group members. In the water sector, forms of self-regulation have been practiced by water service providers for decades through their utility or professional associations, although I doubt that those involved in the early years would have recognised it as self-regulation. Generally this has been limited to technical matters in setting technical standards and codes of practice, although in recent years it has been extended to the establishment and management of benchmarking schemes, as in the Netherlands and Australia. In the Netherlands, the public sector water service provider managers see it as part of a defence against privatisation and associated independent regulation. I am not aware of any countries where water charges are self-regulated by utilities through an association.

Self-regulation can be introduced as part of seeking recognition and respectability. In Ghana, private water tanker operators have formed associations in which members operate to a Code of Practice. This Code has been developed with the regulator (the Public Utilities Regulatory Commission), but is enforced by the association members themselves. This should provide benefits to the communities served by the tankers, providing for discipline on charging and the prospect of better control of water quality. The benefits to the private operators should be that they have become recognised as an important part of the public water supply distribution system, and should be in a better position to negotiate with Ghana Water Company on water charges.

With the benefits of low cost, and with good knowledge of costs and investment requirements within utilities, why shouldn't self-regulation be the norm? After all, it is the form of regulation usually practiced by local authorities. Unfortunately, it has generally failed to produce efficient and sustainable water services, due to water charges being set for political expediency rather than being determined objectively.

WHY INDEPENDENT ECONOMIC REGULATION?

Essentially, due to the high cost of the required infrastructure, particularly water mains and sewerage pipe networks, water supply and sanitation services are monopolies. Some parts of the UK have introduced or are introducing direct competition for non-household customers using common carriage, which is discussed in Chapter 10, but this chapter assumes that there is a need to manage a monopoly situation. Contracting out water service management to the private sector provides for the occasional element of competition, at the time of the initial tender, and subsequently at the end of each concession, but this is infrequent and the choice rests with the asset owners, not with the consumers. Contracting out has introduced the term "regulation by contract". There are differing views as to whether this is regulation and it is discussed later in the chapter. Consumers cannot rely on the market to protect them from higher than necessary charges, from poorer than required quality of service, or from quality of water that does not meet specified standards. So service quality and "product" quality have to be defined, and the price for their delivery has to be determined through some other mechanism.

Generally, water services are provided by local government, with charges fixed internally by those local authority suppliers, either in a recognisable water service bill, or as part of local taxation. This is described as self-regulation, and it is argued that this is an effective method, because if the responsible officials have been elected, they have a duty to serve their constituents and failure to do so will result in their removal at the next election. However, because budget determination and price fixing are internal processes, and only one part of a much bigger local authority portfolio of functions, there is little transparency, so consumers do not have an understanding of the relationship between water charges and the cost of providing the services. As a consequence of this, and with little direct public participation, increases in water service charges under this self-regulatory regime are not only unpopular, but also viewed with suspicion. So, instead of overcharging, local political pressures have tended to result in undercharging, with water service prices set too low for sustainable cost recovery. The situation is compounded in situations where water income is used to subsidise other local services, leaving even less money for maintenance and refurbishment of the infrastructure. There are exceptions to this gloomy picture, with some very good local authorities operating self-regulation in an effective and transparent way. These exceptional examples are described in Chapter 7. Even in these cases, changes in local political leadership could result in a change in policies and priorities with a consequent risk to performance of water services. fu Chapter 1, as a major policy consideration, I argued the need for separation between policy, regulation and delivery functions. Therefore, although I do not wholly disregard self-regulation, because there are some situations, such as in Singapore, and in cities like Seattle in the United States, where it has worked well, I believe that generally there is a need for some form of independent and objective regulation. The need for separation of policy and delivery is becoming increasingly recognised [10], and public sector reform most commonly involves the establishment of "corporatised public" utilities. The approach taken to this reform, with successful examples, is given in Chapter 7. This separation creates the opportunity for the employment of good operational managers, with objectives focussed on efficient service delivery. It also provides for separate accounting and greater transparency. However, if price fixing, or price approval, continues to rest alongside the policy function, there remains the risk of political expediency in the determination of charges. The need to separate the regulation function from the policy making body has not yet received universal recognition, but to me it is essential. An independent body is necessary to give objectivity to the process of turning policies into reality, so that unrealistic timescales for the ambitions of politicians become identified through the planning and price determination process. This is an appropriate juncture to consider approaches to setting tariffs.

APPROACHES TO SETTING TARIFFS

The regulator, whom economists [11] consider a welfare maximiser and the utility (which if private is considered a profit maximiser) have important constraints. The utility as a monopoly is required to meet all demand at the regulated prices and is restricted in its level of profits. The regulator whilst protecting consumers has to ensure that the utility can finance its commitments. This is a fine balancing act between being too soft on the utilities resulting in higher tariffs and being too tough which impacts on utilities' ability to raise capital for the required infrastructure investment. There are a number of models which are used. To a certain extent the choice for regulation depends upon the ability of the regulator to obtain good data comparable with that of the utilities. Six models are discussed briefly; iterative, yardstick, rate of return, price-cap, revenue-cap and efficient firm.

An iterative process is useful where a regulator has little information. The regulator sets prices based on the period before. The profit levels and supply volumes for the period then provide the basis for prices for the next period and so on. This process continues until the allowable prices just cover costs based on the previous period. This can be a slow process, and it is difficult for the regulator to take into account the requirements of the poor. Each utility is considered without reference to any other.

In the yardstick model, the regulator uses information on one utility to regulate the others in the group of utilities. The "chosen" utility is then regulated on the basis of information from the other utilities. One argument for this approach is that the utilities do not have an incentive to give false information as their situation is not influenced by their own data. An advantage for the regulator is that this approach requires a low level of information. For this approach to work there has to be a number of utilities with similar demand and cost conditions. To maintain this condition, regulators tend to oppose mergers or takeovers.

With a rate of return approach utilities are allowed to charge prices equivalent to a competitive market. The regulatory aim is to set these charges equal to efficient costs of water services operations plus a market determined rate of return on capital employed. This approach was used almost universally until the 1990s but it is now recognised that it can result in "cost-padding". Also it is criticised as encouraging use of capital rather than labour, although this charge can equally be levelled at some other approaches in which cost of capital is a critical factor in the tariff calculations.

In the price-cap approach the regulator calculates limits (caps) on tariffs using the formula CPI-X (ie Consumer Price Index less an efficiency factor). X is dependent not only on the past performance of the utility, but also on the performance of the other utilities in the group being regulated. The use of the X factor is considered a proxy for the market. The regulator has to avoid using "acceptable rates of return" in the X calculation to prevent it becoming rate of return regulation. Price-caps as used in England and Wales are described in Chapter 10.

Revenue cap is similar to price-cap but with the cap on total revenue rather than prices. Less used than price-cap but useful for sectors (such as electricity) where possibly the utility can have less influence on the demand, and where costs are independent of short-term variations in demand. Both price-cap and revenue-cap approaches provide incentives for companies to outperform, as they can be set up to allow utilities to retain additional profits until the next price review.

Efficient firm regulation is similar to yardstick regulation but uses the concept of the "perfect utility" to identify opportunities for efficiency improvements. It involves setting prices equivalent to long-term average costs. This is optimal when the utility is required to be self-financing. It requires the regulator to have precise information in order to define the perfect utility. Its use in Chile is described in Chapter 9. The perfect utility concept can be used by utilities themselves to set their own performance targets. Dunea, the water utility in The Netherlands serving The Hague, estimates how much cost saving a perfect Dunea could achieve and sets targets accordingly.

One preferred approach is some form of CPI-X formula to determine price caps, but arguments between economists on the relative effectiveness of approaches continue. CPI-X has been described (12) as "rate of return regulation with a lag". Nevertheless the system in England and Wales can be considered to have delivered a major investment programme without excessive increases in charges. To my mind the choice of tariff calculation formula may be less important than a periodic planning system which provides for all requirements to be considered in an integrated way. Associated with this is the important question of how to determine investment programmes to refurbish neglected underground infrastructure. This is discussed in Chapter 3.

WHY INDEPENDENT DRINKING WATER QUALITY REGULATION?

The responsibility for regulating drinking water quality often rests with those local authorities who are also responsible for the water service delivery, so in these circumstances there is self-regulation. This can be very effective, with those responsible for monitoring and reporting drinking water quality being dedicated people with high integrity. Difficulties can arise when there are competing demands for investment to meet standards, or for example if a local politician would rather that a report on a water quality incident was not made public just before an election. So the arguments made above for independence for economic regulation apply equally to drinking water quality regulation. Above all, an independent regulator can ensure that there is transparency in the reporting on drinking water quality and on the investigation of incidents affecting drinking water. It also provides an independent audit function necessary for the attainment and maintenance of high professional standards within water supply operators. Effective drinking water quality regulation is covered in Chapter 4.

ENVIRONMENTAL QUALITY REGULATION

It is accepted generally that the regulation of the environment is a separate and independent role from that of water supply, because the function goes beyond the water supply chain. The roles were combined in England and Wales during the period 1974–1989 (see Chapter 9) in the operation of the Regional Water Authorities. Although seen by some as utopia at the time, the results were not as hoped for, partly due to financial constraints but probably also due to a conflict of objectives when considering water resource needs. A similar conflict of objectives exists when an environment regulator has responsibility for both environment protection and water resources. Clearly, decisions on new water resources, whether involving new reservoirs, new boreholes, or new river abstractions, have to be considered in the context of the impact on the environment, and the decision process has to balance people's needs against ecological judgements. This difficulty can be compounded by environmental pressure groups generally being better organised than the general public, and being in a better position to influence an environment regulator's decisions. To ensure objective decisions there is a need for a means of obtaining equal input from all public viewpoints and a means of independent review. However, this problem is minor compared with the need for regulation to be independent and separate from government, to avoid the risk of decisions being made "behind closed doors" for political expediency.

REGULATION BY CONTRACT

I begin by stating my concern over the limitations of regulation by contract. Clearly a well-defined contract provides the means of controlling the delivery of the desired performance of a water service operator, whether public or private. That is a major step forward and much better than the "traditional" municipal operation in which there are limited or no performance measures and poor financial ring fencing. What it doesn't do is provide any means of regulating the contract owner, nor does it provide for transparency and public participation. On the contrary, it tends to reduce transparency and assumes that governments can look after the consumers' interests. Also, this lack of transparency risks providing the conditions for corruption. Regulating the contract owner might be considered a strange concept, but over the period 2001–10 there was a good example of this in the United Kingdom. Dŵr Cymru, the now "public" licenced "water undertaker" providing water services to Wales, chose to contract out the delivery of its service to private companies. It controlled that delivery through the conditions in the contracts, but Dŵr Cymru itself was subject to regulation by the economic regulator Ofwat. Any penalties for failures to meet regulations fell on Dŵr Cymru, not on the contractors. The contracts can be subject to commercial confidentiality without fear that the consumers will suffer, because the regulatory process ensures transparency in the process of determining tariffs and in delivery of the service commitments. I appreciate that this was a unique situation, but to me it provided a way forward on the regulation of water services under Public Private Partnership situations. It means that should be some independent regulation dealing with the objectives and deliverables around a contract, assuring transparency on performance. This should not interfere with the contract itself, which should be devoted to achieving the outputs at a competitive price. The Dŵr Cymru example is also relevant to the recommendation in the UNHD Report 2006 already mentioned, which advocates some form of regulation "... to ensure accountability to the poor", as taking this forward would involve regulation of local government.

There have been too many contract failures to assume that it can be left to "regulation by contract". In Chapter 8, on the successful use of the private sector, I give some examples of both successes and failures. The reasons for failures can be attributed as much, or more, to the contract owners as to the contractors. What is often missing is an effective planning process which considers objectives, costs and charges in an integrated way. If this process is handled by an independent regulator, unrealistic political ambitions, such as solving infrastructure problems at the same time as cutting overall costs whilst providing additional income for a city, will become apparent. Once realistic deliverables have been established and defined clearly, a contract can be placed by the local government for delivery. The cost of the contract together with other costs for sustainability of the water service can be incorporated into charges, and in subsidy provisions, in a transparent way. There is also the need to have a means of handling changing circumstances, whether they are political, legal or physical, and this requires a periodic review. This needs to take place at least every five years, and is best handled by a regulator. The contract needs to include provision for such reviews, and provide the basis for the contract conditions to be adjusted to reflect the changed delivery requirements, with a corresponding price adjustment. This process should avoid "fudging" (i.e. "glossing" over, or ignoring) issues for political expediency, provide contractors with a stable commercial environment, and also give them respectability. A regulator is in a position to achieve necessary public participation in a way which will focus on people's needs and on the concerns they have for affordability, whilst avoiding the whole focus being on the emotive issues associated with the use of the private sector. So, to my mind, a contract provides for a sound basis for monitoring delivery, whether by a private or public service provider, but it is not sufficient for regulation, which needs to include the process for determining realistic objectives, how costs will be recovered for sustainability, how changing conditions will be accommodated, and the involvement of the public.

The World Bank Explanatory Notes [2] warn about the problems of mixing the two "traditional" designs of regulation, referring to them as the French contract-based tradition and Anglo-American style regulation, and give examples from around the world where hybrids have failed. Considering those cases, the failures appear to have been due to interference by the regulator in the contract details. As I have explained above, I don't see the regulator as interfering with the contracts, but providing the objectivity and transparency on which the contracts can be based. The Notes, in considering how there may be benefit in involvement of a regulator, refer to "creating an autonomous body to monitor the contract, enforce it, and provide the analytical input when tariffs or other aspects of the contract are reset or renegotiated". This is consistent with the process I have outlined in the previous paragraph. Concern should go beyond having efficient delivery in the short term, to having sustainable water services in the longer term, and this means sustainable cost recovery. History tells us that governments, without the benefit of an independent regulator, do not find it possible to set, or maintain, a level of water charges necessary to achieve sustainable cost recovery. However, there is a tendency to increase charges at the time of awarding a concession contract, to pay for the cost of service improvements built into a contract. This leads to public resistance to the use of the private sector, with the risk that pressure groups, opposition parties and the media will look for any opportunity to undermine any ongoing commercial commitments. Independent regulation, with effective public participation, can alleviate such a situation, through providing a climate of better public understanding and trust, which regulation by contract is unable to do.

In the developed world today, waterborne-induced illness is rare and there are few deaths from waterborne disease, but this was not the case before water treatment and safe handling of human excrement was introduced at the end of the 19th century. The cholera outbreak in London of 1854, with the pioneering epidemiological work of Dr John Snow, was an important step in the understanding of waterborne disease. A book [12] by Steven Johnson adds to the information on the arguments about the nature of the transmission of infection at the time of the outbreak. However, in the developing world today, even though it is understood how and why, poor access to safe water and inadequate sanitation is a major cause of illness and deaths. According to Global Health Observatory, WHO, in its report of 2004, there were 1.9 million deaths per year from diarrhoea of which 88% were attributable to unsafe water, inadequate sanitation and lack of hygiene [13]. Higher and lower numbers are quoted. Whatever the true figure, waterborne disease results in a large number of deaths each year. Those most affected are children under the age of five in developing countries.

Standards need to be designed to meet citizens' requirements relating to drinking water safety and aesthetic quality, taking into account logistics and affordability. In the developed world there is a tendency now to make standards unnecessarily stringent, and in the developing world to adopt standards that are unrealistic in the short to medium term. Almost everywhere too little attention is given to monitoring and enforcement.

I stress at the beginning that there is no point in having standards unless they are monitored, there is no point in monitoring compliance with standards unless it is followed by enforcement, and no point in having an enforcement process unless there is a means by which the necessary improvements can be funded and implemented. This chapter tracks the steps from threats to health leading to standards, to current thinking on how best to achieve safe drinking water, to monitoring and enforcement, to sanctions for failures. The chapter finishes with the example of the Drinking Water Inspectorate in London as one good model of a drinking water quality regulatory body.

THREATS TO THE SAFETY OF DRINKING WATER

The threats can be considered under three headings, namely microbiological organisms, chemical substances and radionuclides, with pathogenic microbiological organisms being by far the most important. It is difficult to give accurate figures on numbers of cases of illness and deaths from pathogens but estimates give cases in billions and deaths in millions each year. Figures taken from various WHO documents are given in Table A.1. There is a tendency by lay people to see the main threat as cholera, presumably because such outbreaks hit the headlines. The figures for cholera vary considerably year by year (Figure A.2) but its impact although important is only part of the diarrhoeal disease total

Table A.1 Incidence and deaths from waterborne disease.

Water, sanitation and hygiene related pathogens		
Disease	**Cases per year (thousands)**	**Deaths per year (thousands)**
Total Diarrhoeal Disease	2,000,000	2.400
Cholera	3000–5000	100–120
Giardiasis	2000	low

Sources: GBD2004 and CDC.

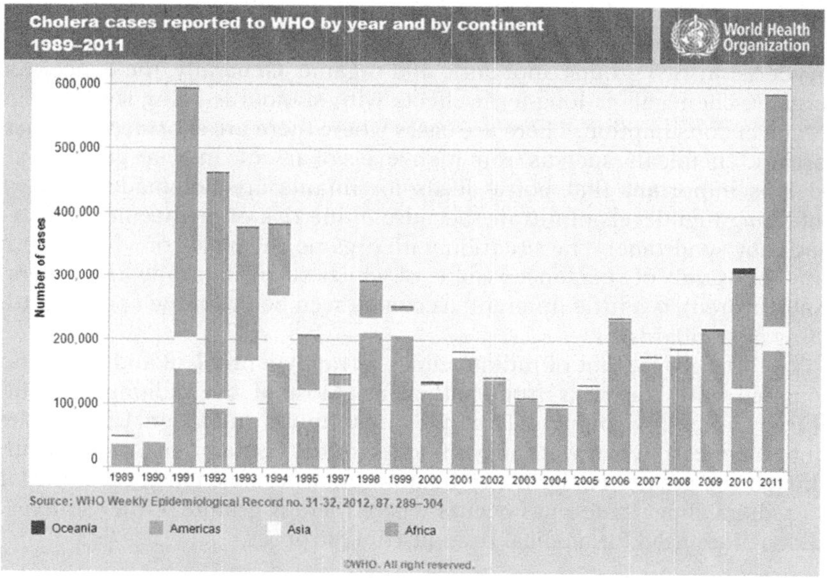

Figure A.2 Number of cholera cases reported from 1989–2011 (*Source*: WHO).

from many different organisms. Table A.1 includes the number of cases of the illness giardiasis. It is estimated [14] that around one third of the population in the developing world has been affected by giardia. It is the most common cause of waterborne illness in the United States [15], but there are few deaths. In the developed world protection against *giardia* and *cryptosporidium* has become a major consideration due to these organisms being resistant to chlorine.

The Global Burden of Disease Study, GBD10, [16] published in December 2012, described as the largest systematic effort on the subject, compared the situation in year 2010 with that in 1990. The study concluded "The results show that infectious diseases, maternal and child illness, and malnutrition now cause fewer deaths and less illness than they did twenty years ago". Table A.2 gives the figures from that study on deaths from total diarrheoal diseases, cholera and cryptosporidiosis. At the time of writing the data and the methodology used in GBDlO are being debated.

Table A.2 Comparison of deaths in 1990 and 2010.

Estimated worldwide deaths 1990 and 2010 (thousands)		
Disease	**1990**	**2010**
Total Diarrhoeal Disease	2487	1445
Cholera	120	58
Cryptosporidiosis	222	99

Source: GBD10.

In comparison with microbiological organisms, there is less risk from chemical substances which occur rarely at toxic levels. Chemicals are considered in two groups, inorganic and organic. Generally, the concern with inorganic chemicals is long-term effects with standards being set based on a lifetime of consumption. There are cases where there are short-term impacts of inorganic chemicals, such as from high levels of arsenic in some groundwaters, and it is important that bottle feeds for infants are not made from water containing high levels of nitrate, because of the risk of methaemoglobinaemia (blue baby syndrome). The situation with organic chemicals of which there are many thousands of substances (some of which occur naturally and others are manufactured), is a little different, as can be seen below in the approach to the setting of standards.

The third group, that of radionuclides, is rarely a problem and insignificant compared with the risks from pathogens. Most of the radionuclides occur naturally in the environment. Examples are uranium and radon. Uranium levels can be a concern when water from flooded mines contaminates water courses, although the greater concern can be uranium as a toxic chemical rather than as a radionuclide. Radon gas occurs where there is granite. Other sources are radionuclides used for medical or industrial purposes.

APPROACHES TO THE PROVISION OF SAFE DRINKING WATER

There can be considered to be two approaches, the setting and monitoring of standards, and risk management. In practice the two approaches go together, although recently there has been a change of emphasis towards effective risk management.

Standards can take two forms, numerical and treatment. Numerical standards, which are levels of substances which should not be exceeded, are set for actual contaminants. There are also indicators which are important as a means of determining whether one or a number of undesirable "things" are present. An example is *e-coli* which, although not a pathogen itself, it is an indicator of faecal contamination in which certain pathogenic organisms could be present. The presence of *e-coli* indicates the need for a thorough investigation.

The requirement to disinfect water intended for human consumption is one example of a treatment standard. Treatment standards can be based on the effectiveness of specific treatment processes, and can be expressed in numbers of log (logarithmic) removal. A one log removal process will reduce the quantity of "a particular thing" to one tenth, two log removal to one hundredth, three log removal to one thousandth, four log removal to one ten thousandth, and so on. Setting standards on this basis requires an assumption to be made that the numbers of something remaining will not constitute an infective dose. There can be alternatives of a treatment standard or a numerical standard, as in the UK approach which was used to protect against the organism *cryptosporidium*. This is discussed later in this chapter.

The recent change to give greater emphasis to risk management arose from concern over the tendency for more and more numerical standards. For example, in the United States, there was a legal requirement to include additional substances periodically. A number of people closely involved in drinking water quality said that there must be a better way of approaching drinking water quality, because although numerical standards are important, measurement of quality at the point of delivery is too late, other than for verification. As a result of this concern, the first of two very important international workshops was held in Bonn, Germany in October 2001. At this workshop there was a fundamental review of the approach to the delivery of safe drinking water. This led to a change of emphasis in the revision and development of The WHO Guidelines for Drinking-Water Quality. There was a second Bonn workshop in February 2004 which led to the Borm Charter. The Bonn Charter [17] and the WHO Guidelines [18] are interrelated and complementary documents. They were launched together in September 2004. They both refer to the need for "water safety plans" as a disciplined approach to risk management. Although I prefer the term "drinking water safety plan", as it makes it clear that the reference is to drinking water, in this book I use the term "water safety plan" for consistency with WHO terminology. I will return to setting numerical standards later in the chapter.

THE BONN CHARTER

The Bonn Charter is a set of high-level principles which can be applied worldwide for the supply of safe drinking water. It is important to recognise that the Charter is concerned with drinking water quality, although many of the principles apply equally to all aspects of water supply. Figure A.3 gives a diagrammatic representation of the Borm Charter framework. The important goal is described as "Good safe drinking water which has the trust of consumers". There are some key words here. "Safe" means safe, especially safe from pathogens. The interpretation of "good" has to take into account in which country the definition is being applied as the aesthetic quality depends critically on affordability. The word affordability does not appear in the Charter, although there is reference to the need for standards to reflect the progressive realisation of goals. I make reference to affordability, as we have to consider the world as it is, not as ideally we would wish it to be. In this context, I discuss below the need for interim achievable standards. Returning to the goal, the word "trust" is important, reflecting the need for good public participation, and the importance of full reporting of drinking water quality.

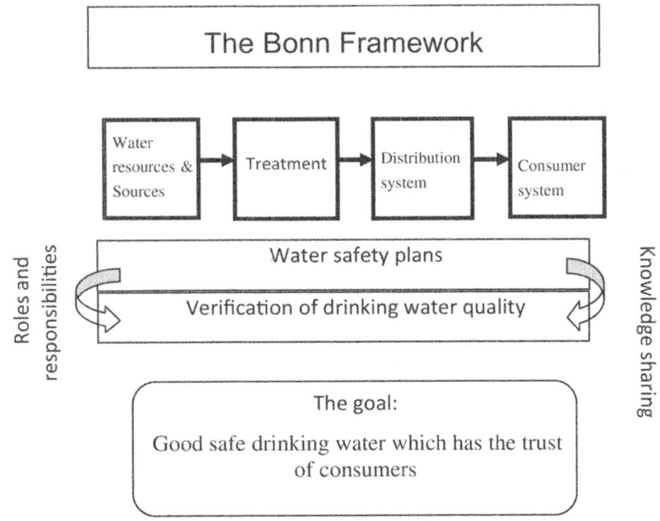

Figure A.3 The Bonn Framework (*Source*: The Bonn Charter).

The Charter aims to provide the various stakeholders, politicians, government officials, regulators and senior water utility managers, with a framework for the development of an integrated approach to the delivery of safe drinking water, giving some key principles and listing roles and responsibilities. Some of the key principles are:

- there is a need for an integrated approach to include all organisations involved from catchment to tap;
- there should be close co-operation between all in the supply chain;
- transparency is vital for consumer confidence;
- drinking water quality standards, at consumers' taps, will depend upon local circumstances, but should cover both safety and acceptability;
- the need for clarity of responsibilities for each part of the supply chain.

The Charter considers the roles and responsibilities of the various stakeholders. Among the key responsibilities for governments are the creation of independent systems of coordinated water quality regulation, and the provision for institutions and coordination arrangements to deal with the risks to water quality from catchment to tap. Key responsibilities of water service providers, the utilities, include taking the lead, and working with others in the development and implementation of water safety plans. The development of water safety plans is covered in a section below. Another responsibility of utilities is ensuring that the full cost of service provision is identified, including the cost of maintenance and provision of assets. This is another context in which the need for full-cost recovery is important. In addition to the core responsibility of regulators in relation to drinking water quality standards, stress is given to the need for transparency and the full reporting of compliance results in a way which can be understood by all. Consumers themselves are regarded as having responsibilities in preventing contamination of drinking water in houses, including avoiding the use of inappropriate plumbing materials. Importantly, the Charter implores all parties to share information pertinent to achieving safe drinking water.

WHO GUIDELINES

At this stage it is appropriate to show the relationship between the Bonn Charter and the WHO Guidelines, which is illustrated by a combination of Figures A.4 and A.5. Figure A.4, taken from the WHO Guidelines, gives the structure of the guidelines, showing the emphasis given to "water safety plans" within the central framework for safe drinking water. The guideline values

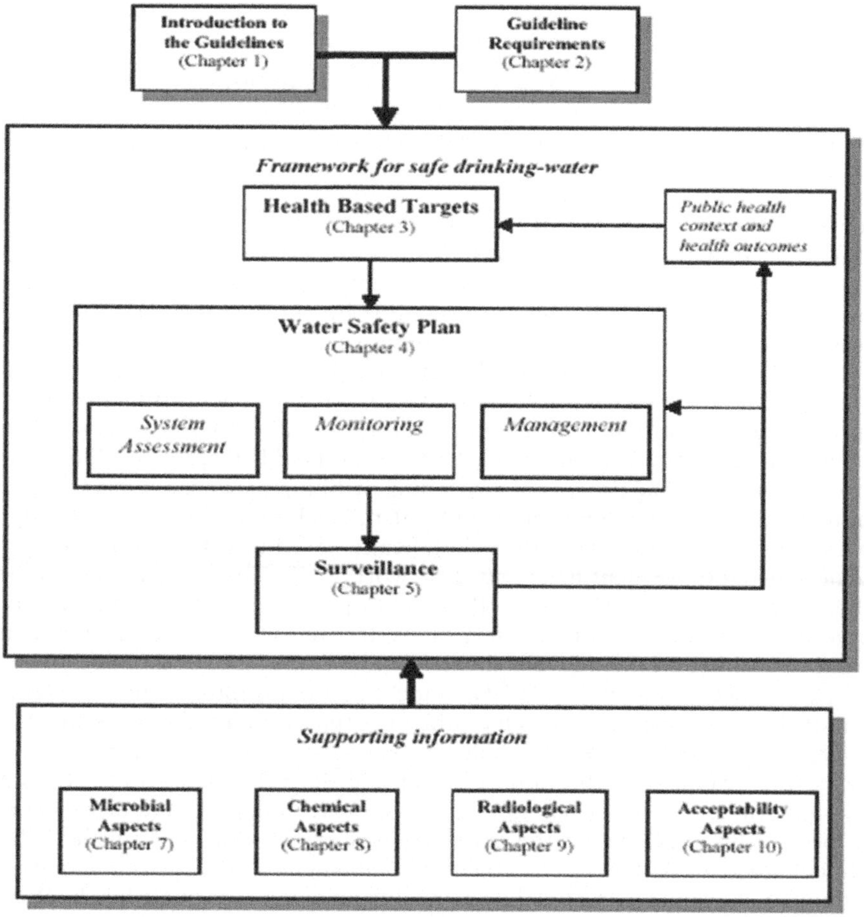

Figure A.4 WHO Guidelines Structure (*Source*: Guidelines for Drinking Water Quality, WHO 2004).

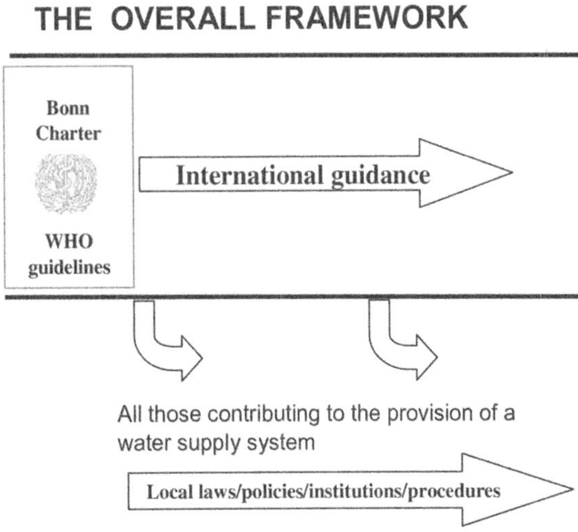

Figure A.5 Integrated International Guidance (*Source*: Presentation at Launch of Bonn Charter 2004).

for microbiological, chemical, radiological and aesthetic aspects, on which standards are based, are now shown as supporting information.

The approach includes the setting of health-based targets. Although this is highly desirable, with the aim of setting investment in water services against health needs, and measuring the health improvement impact of that investment, I have concerns about the practicality of the approach, especially in developing countries. It is known from a WHO study [19] that the benefits from investment in water and sanitation in developing countries greatly exceed the costs. Also, obtaining adequate data on which to base the health targets is difficult and time consuming. The process of developing drinking water safety plans, which in many cases can be done very simply, identifies both investment and system management priorities and, in my view, health-based targets are not necessary to decide that those identified actions are worthwhile. There is also the risk that the effort taken in producing health-based targets could delay the preparation and implementation of water safety plans or, even worse, deter governments from embarking on their development.

The Bonn Charter and the WHO Guidelines provide international guidance, but their implementation involves the incorporation of the recommended approaches, together with the guideline values, into local laws, policies and procedures. With both the Bonn Charter and the WHO Guidelines stressing the importance of water safety plans (WSPs), it is appropriate at this stage to begin with guidance on the preparation of plans. Chapter 4 of the WHO Guidelines gives detailed advice on the development of plans with reference to a number of examples. A book [20] providing guidance was published by WHO in 2005. Since that time WSPs have been prepared in many countries across the world and extensive guidance has become available, much of which is incorporated in the Water Safety Portal [21]. In 2012 WHO published guidance on water safety planning for small community water supplies [22]. The Water Safety Portal provides guidance on preparing WSPs, provides tools to assist in their development and gives some case studies, and provides support to Water Safety Plan Networks covering Africa, and the Asian-Pacific and Latin American-Caribbean regions. The Portal gives 11 modules in the preparation and monitoring/audit of water safety plans. They are:

- Assemble the WSP Team
- Describe the water supply system
- Identify the hazards and hazardous events and assess the risks
- Determine and validate control measures, reassess and prioritize risks
- Develop, implement and maintain an improvement/upgrade plan
- Define monitoring of the control measures
- Verify the effectiveness of the WSP
- Prepare management procedures
- Develop supporting programmes
- Plan and carry out a periodic review of the WSP
- Revise the WSP following an incident

WATER SAFETY PLANS

Readers wishing to embark on the preparation and implementation of WSPs should study the advice and experience available on the Portal and elsewhere. Based on my own experience I discuss some of the steps, consider audit of WSPs, and refer to WSPs in developing countries including their use as a means of informing improvement programmes.

The boundaries of a system can depend upon the source, the area supplied by a treatment works, or the system supplying a defined distribution system. The choice can be for convenience, or because a particular supply system is susceptible to some risks of known concern, or to reflect a team's responsibilities. A source can be a reservoir or river catchment, an aquifer, or a desalination plant, or a reuse source or, as in many cases, combinations of individual sources. Based on the choice and definition of the system being covered, a project team is formed. The team should include members from each organisation in the supply chain covering the water source, water treatment, water distribution and, if possible, consumers' premises. Generally, the team should include key people from a number of different organisations; this requires the management of those organisations to recognise the importance of collaboration and sharing of information. Drinking water safety is far too important for turf wars to get in the way of teamwork, not only in the development and operation of water safety plans, but also in the operation of systems in the future. The team leader is most likely to be a member from the water service provider, but the choice can depend upon circumstances, and on the availability of key individuals.

A good starting point for the team is an initiation workshop facilitated by someone with experience in the process of producing water safety plans. A workshop provides the opportunity for the team members to get to know each other, and to understand each other's challenges. It is desirable to invite others, who can provide valuable inputs on the supply history and potential risks to supply, and for there to be management involvement to ensure a commitment to implementation. At this stage it is important to assemble available information and documents on the system under study, such as an accurate map and flow diagram of the treatment works, and to identify deficiencies in any of the information. One output from the workshop should be an agreed timetable for the project, with a firm date for the completion of the frrst (risk assessment) stage.

The risk assessment involves each member of the team identifying the hazards to the safety of drinking water in his or her part of the supply chain. Hazards can be from a number of sources: environmental including pollution incidents, system failure, or from deliberate attempts at contamination by man. Dealing with risks from sabotage, terrorism or vandals is often seen as a separate requirement, but I consider it important for these to be included in one drinking water safety plan, as all aspects need to be built into operational procedure and reaction planning. At this stage I have used the words "hazard" and "risk" rather loosely. The Oxford English Dictionary does not help very much but there is a useful discussion [23] of the distinction between the two words by the UK Centre for Occupational and Environmental Health. COEH refers to hazard as the potential to cause harm, and risk as the likelihood of harm (in defined circumstances, and usually qualified by some statement of the severity of the harm). This highlights two important measures, likelihood and severity, which need to be considered together, and the WHO Guidelines provide a simple scoring system (Table A.3) to assist in identifying the relative importance of risks. Such a simple system should be used only as a guide because, for example, the catastrophic nature of a hazard may be such that, however unlikely it is to happen, it needs to be taken very seriously. The risk-scoring table uses the following definitions which are useful, but again they should be regarded only as giving some guidance and not used prescriptively:

Likelihood

- Almost certain – once a day
- Likely – once a week
- Moderate – once a month
- Unlikely – once a year
- Rare – once every five years

Consequences

- Catastrophic – potentially lethal to a large population
- Major – potentially lethal to a small population
- Moderate – potentially harmful to a large population
- Minor – potentially harmful to a small population
- Insignificant – no impact or not detectable

Each hazard and its risk assessment should be recorded, together with any supporting information. Table A.4 is an example of a recording system.

Once the risk assessment has been completed, each member of the team should review the controls in place to counter the hazard. Is there evidence

Table A.3 WHO suggested risk scoring; a simple risk scoring system.

likelihood	Severity of consequences				
	Insignificant	Minor	Moderate	Major	Catastrophic
Almost certain	5	10	15	20	25
Likely	4	8	12	16	20
Moderate	3	6	9	12	16
Unlikely	2	4	6	8	10
Rare	1	2	3	4	5

Source: Water Safety Plans The World Health Organization 2005.

that the controls are working? Have there been incidents where the controls have been inadequate? What preventative measures could be taken to reduce the risk? Could improved or additional downstream (i.e. later on in the supply chain) controls be put in place? To give an example, in the case of a hazard of a river pollution incident, there should be preventative measures, such as bunds on a factory premises to contain a chemical spill. A downstream measure could be detection of pollution in the river, coupled with the closing of the intake to the treatment works and, if necessary, additional treatment at the works. In another case, a downstream control for a failure of disinfection at a treatment works could be chlorine injection into the distribution system. The analysis of control measures is likely to identify some priorities for improvement which should be acted on at this early stage, thus giving some early benefit for the work done on the project. Although the control improvements will reduce the likelihood of an incident taking place, they will not change the consequences should it do so.

The next step is to set both target levels and action limits for each of the controls, and to record the designated reaction measures should an action limit be breached. The target levels on controls at a treatment works are likely to include the drinking water quality standards at which, in general, noncompliance can be dealt with under routine conditions. An action level should be set at a figure at which immediate action needs to be taken. At the beginning of the supply chain, at the source end, some of the target levels are likely to be more subjective and could require longer timescales for achievement. For example, it may require a 10-year or more programme to “clean up” a catchment. In such cases, the action limits need to be set to give as early a warning as possible for downstream reaction programmes to be enacted. Monitoring against the limits is critical, and it is necessary to define what, when (frequency) and by whom. The other necessary component is a means of verifying that the corrective action has worked. The outcome of this work can be recorded as in Table A.5 below.

The above provisions should be converted into operational procedures, then operators should be trained in the new or revised procedures; the management responsible for each part of the supply chain should establish a means of checking that the procedures and training do result in an effective system for the supply of safe drinking water supply. Exercises which try to simulate real incidents provide a useful means of checking the effectiveness of the

Table A.4 Table for recording risk assessment information.

Part of supply chain	Source/ cause	Hazard	Preventative and control measures	Likelihood	Severity	Risk	Downstream controls	Comments and controls further considerations

Table A.5 Record of action limits, monitoring and corrective actions.

Hazard	Critical limits		Monitoring			Reaction steps	Verification
	Target level	Action level	What	When	Who		

systems, especially the critical aspect of communication between the various organisations involved from catchment to tap.

BEYOND THE IMMEDIATE CONTROL OF THE WATER SERVICE PROVIDER

The Bonn Charter makes it clear that hazards have to be considered from source to consumer. Some of the greatest risks may exist once the water has left the water service provider system whether that be a piped system or a well. In many parts of the world final distribution relies on the informal sector delivering water by road tanker or by cart. Water may have to be stored in tanks or multiple containers. Such water can be vulnerable to contamination. WHO provides advice on protecting drinking water in households [24]. Devices for treating water in premises are available but these present hazards if they are not maintained. WHO and UNICEF have published a toolkit [25] for monitoring and evaluating household treatment and safe storage programmes. The toolkit refers to water safety plans as a useful instrument in determining when household systems are appropriate. The importance of hygiene measures such as hand washing and covering water storage containers is stressed. Consumer and Community Groups have a role in getting people to understand the risks to health and how they can be managed. Advice to householders should be provided in an easily understood form. This may include leaflets giving advice in illustrations for the illiterate. With concern over the risks of cryptosporidiosis to the immune-compromised, WHO issued a bulletin [26], advising on household control measures.

GOVERNANCE AND REGULATIONS

The extent to which the emphasis in the delivery of safe drinking water has moved towards managing risk is reflected in intentions for future regulations, and in how regulators [27] are adopting the thinking. During December 2006/January 2007 the International Water Association (IWA) held an internet forum on the Bonn Charter; those taking part described how the Charter and drinking water safety plans were being taken forward in a variety of ways, depending on countries' approaches to water management, standards and regulation. There were indications by the European Commission that water safety plans should be encouraged, or even incorporated, in the next EU Drinking Water Directive.

Increasingly water safety planning is being adopted in policy and embedded in regulations. I give a few examples.

The New Zealand Health (Drinking Water) Act 2007 requires the establishment of public health management plans for all supplies whether urban or rural. Supply units serving less than 5000 people can obtain training and assistance in writing plans through the Technical Assistance Programme. Assessors appointed by the Director-General of Health check whether suppliers are in compliance with the Act. There is enforcement via an escalating series of penalties for non-compliance.

The Singapore drinking water regulations, the Environmental Public Health (Quality of Piped Drinking Water) Regulations, require water suppliers to develop and operate water safety plans to a specified standard. The requirements are set sown in the Code of Practice on Piped Drinking Water Sampling and Safety Plans. The Drinking Water Inspectorate audits the plans and their operation.

The regulator in Ghana, the Public Utilities Regulatory Commission (PURC), responsible for drinking water quality piloted an approach to water safety planning and provided initial training for the roll out of WSPs across urban supplies. Procedures for the audit of the plans are built into the PURC Inspectors' Manual. The pilot project is discussed later in this chapter.

In South Africa [28], as part of drinking water quality incentive-based regulation, water safety and incident response management plans are required. The certification system, named blue-drop, specifies risk assessments of catchment, treatment works and reticulation systems. A paper [29] published in 2006 had provided the framework for a risk management approach.

In India policy issues and governance aspects [30] were considered in the development and application of water safety plans for rural areas. The work addressed a number of key elements for successful application including essential technical support to the Gran Panchayats (GPs, the local parliaments) and their Village Water and Sanitation Committees (VWSCs) who are responsible for the water services. WSPs were seen as an important vehicle in developing improvement programmes, with the need to integrate them into the planning processes. Each layer of government has a role to play to support local operations, as demonstrated by Figure A.6, the inverse pyramid taken from the Water and Sanitation Programme (WSP) document.

The drinking water regulations in England and Wales make specific provisions for drinking water safety and require water companies to implement a risk (water safety plan) approach to water production and distribution. The regulations are enforced by the Drinking Water Inspectorate [31]. The Scottish Water (Objectives for 1st April 2010 to 31st March 2015) Directions 2009 [32] issued by Scottish Ministers required the water company supplying the whole of Scotland to prepare water safety plans for all supplies during the objective period.

In Hungary [33], by Government Decree 201/2001 which came into force in 2009, all water supplies serving more than 5000 people are required to have water safety plans with progressive implementation by July 2014. Some emphasis is given to risk management of drinking water resources.

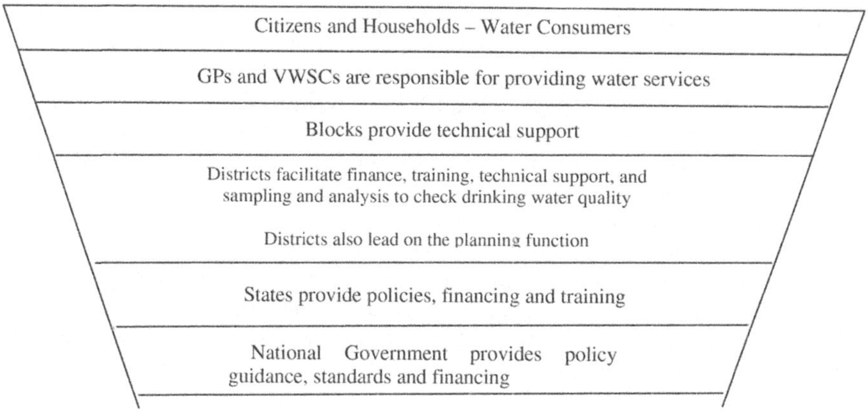

Figure A.6 Inverse Pyramid of Support (*Source*: WSP Delhi).

An excellent example of regional co-operation is in the Pacific Island Countries who came together to develop an approach to drinking water safety through the introduction of sound policies and water safety plans. The resulting report [34] is not only a valuable account of the approach but it also provides a good manual on water safety planning.

AUDIT OF WATER SAFETY PLANS

Audit approaches can be split into three groups, self-audit, third party accreditation, and audit by a regulator. The most common approach in the developed world is a HACCP-type accreditation system. HACCP (Hazard Assessment and Critical Control Points) is used extensively in the food industry and this formal approach to risk assessment has been adopted in Australia. The approach in Australia ([35] and [36]) has been driven by the requirements of the National Health and Medical Research Council Framework for the Management of Drinking Water Quality (DWQMF). These requirements have been incorporated in the Australian Drinking Water Guidelines [37], which are based on a combination of HACCP and ISO 9000. The Framework gives a quality management system to be followed by water utilities using a catchment to tap risk assessment approach. Using approaches developed for the food industry, there is third party accreditation of the operational procedures and, if deemed by the assessor to meet the required accreditation standard, the water utility is granted a HACCP Certificate.

For some systems, for example in Melbourne Australia, the scrutiny of water safety plans has been taken one step further through using a Framework Continuous Improvement Tool (FCIT) [38], which has been developed by the Water Services Association of Australia. FCIT is a tool to judge the robustness of a water safety plan. It involves asking four searching questions on the understanding and application of around 200 different measures. There is a cumulative score, based on the answers to the questions, with a maximum score of 100. The highest scores achieved to date are around 80. Victoria State plans to use this approach for audit of the Drinking Water Quality Guidelines. The aim, certainly in Melbourne, is to move to the ISO 22,000 standard, which is in the same format as ISO 9001 and ISO 14001, and is described as the tool to harmonise all HACCP based standards. Clearly, this approach leads to highly developed management systems, which should have a big impact on the quality of water supply operations, and thus the safety of drinking water.

The Australians, and others who have made so much progress in this direction, are to be applauded, but I have two concerns. One relates to badges; this section has referred to a number of management standards with success resulting in an organisation being awarded a "badge". In the pursuit of badges, it is important to remember that the primary objective is safe drinking water, not management awards. My other concern, and far more important, is that other countries, which are not yet as well developed, could either attempt "to run down this path" before first learning "how to walk", or could be frightened off by the complexity of the approach and shy away from developing basic water safety plans. I will now consider initial approaches in developing countries.

DRINKING WATER SAFETY PLANS IN DEVELOPING COUNTRIES

Whereas in developed countries, water safety plans are a means of refining operational procedures, in the developing world they also provide a valuable tool for raising the awareness of the fundamental causes of water supply problems, and for establishing investment priorities. A detailed case study [39] was carried out in Uganda. This produced recommendations which included attention to leakage control to reduce ingress into the distribution system, and gave greater emphasis to monitoring chlorine residuals rather than microbiological analysis of water samples. The study showed the importance of giving emphasis to operational management and of the need for management controls from catchment to tap. Good progress has been made on the development of water safety plans in rural areas. In Bangladesh, the approach has been to develop plans according to the type of technology used. The approach has been well accepted by the communities who take responsibility for day-to-day monitoring. Examples for dug wells, tube wells, pond sand filters and rainwater harvesting, are available on the International Training Network Centre, Bangladesh University of Engineering and Technology website [40]. I give below an example of implementation in a developing country, based on a pilot project for an urban water supply, in Ghana.

Ghana water company limited

In Ghana, there is one public utility, Ghana Water Company Limited (GWCL) supplying all urban areas. It is regulated on tariffs and drinking water quality by the Public Utilities Regulatory Commission (PURC). Water resources are the responsibility of the Water Resources Commission (WRC) and the Environment Agency is responsible for monitoring the catchments. In April 2004, there was an initiation workshop, organised by PURC, for a pilot project to develop an approach to drinking water safety planning in Ghana. The Weija

Treatment Works supply system, within the Densu Catchment Basin, which supplies part of the capital Accm, was chosen for the pilot project. A project team was formed with members from each of the organisations mentioned above. The team included a member from the Ghana Consumer Association, who was able to provide input on the consumer end of the supply chain, and to stress the importance of an integrated approach to improving the water supply for the benefit of the community. A risk assessment, a review of controls, a review of monitoring requirements, and the establishment of reaction plans and verification measures, were completed by the team in early 2005. There were a number of key recommendations for action, covering each part of the supply chain. On the catchment there was a need to control the amount of nutrients reaching the lake because this was resulting in the formation of algae and difficulties at the treatment plant, with a high likelihood of particle breakthrough. Requirements on the clean-up of the Densu Basin catchment included relocation of informal farming communities, the introduction of good forestry and logging practice, the introduction of better fishing practice, and the control of leachates from waste tips. This will require a long-term programme, and in the shorter-term there is the option of GWCL considering the construction of a floating intake, to allow water to be taken from across an area of the lake, and at depths where there is a lower concentration of algae.

The treatment works was in good condition. Apart from the problems associated with algae removal, there were no aspects requiring significant investment. There were concerns over the unreliability of chlorine supplies and the need for improved monitoring of chlorine dosing.

The situation with the distribution system was very different with the need for replacement or refurbishment of old pipes, the need for replacement of valves and other fittings, and the need for active leakage control. There were no accurate figures on leakage levels but, given the known poor state of the infrastructure, it could have been as high as 50–60%. From the point of view of drinking water safety, the concern about the state of the system related to the inability to maintain 24 hour supplies. futermittent supplies, with no pressure in parts of the system for many hours, create the risk of ingress of contaminated water from the ground which may contain faecal matter. When water begins to flow again in those pipes, the contaminated water is delivered directly to the point of supply to consumers without any opportunity to make it safe, other than by boiling. This risk received a high score due to the likelihood of daily occurrence with the potential to affect a large population. fuvestment in the distribution system to reduce leakage would achieve two things; it would provide perhaps an additional 60% water available for consumers and allow for a 24 hour supply to those consumers connected to the distribution system. Such investment would have far greater value than additional resources or an extension to a treatment works.

It is important at this time to mention the road tanker operations which, in effect, extend the distribution system by serving those people living out of reach of the system, and also those who have infrequent piped supplies. Drinking water can be contaminated when the tankers are being filled, within the tankers if contaminated water has been transported, and at the point of delivery to the consumers. This operation was identified as a significant risk to the safety of drinking water. In Accra, the road tanker operators have formed associations to bring some discipline to the tankering operations. PURC, working with the Tankering Associations and GWCL, has produced Tankering Guidelines [41] which include provision for the periodic disinfection of the tankers, and for random samples to be taken and tested by GWCL. The Guidelines can be considered as a useful model for use in other countries.

The pilot project has been successful in achieving an integrated approach in the consideration of risk from catchment to tap, in the development of water safety plans. The pilot has shown that, although the problems are almost certainly known already by those struggling with them day by day, the key requirements for improvement are highlighted by the water safety plan approach, thus giving the priorities for investment. Above all, the approach developed in Ghana concentrates on a few major risks with associated critical controls, and involves minimum documentation. It is an approach suitable for adoption by other developing countries as a first stage. Once the major risks have been addressed, the assessment approach can be repeated and, over time, operational procedures can be developed to match those in existence in the developed world.

The water safety plan approach provides the basis for audit by quality regulators and, in Ghana, PURC has developed regulatory procedures with emphasis on checking the effectiveness of risk management procedures. Sampling and analysis remain necessary for verification and enforcement of standards. The regulatory skill is in achieving a balance between the two parts of the audit. It is important that regulators, in responding to the change of emphasis from end of pipe testing to checking the effectiveness of water safety plans, do not encroach on the management responsibilities of water utilities.

VERIFICATION AND STANDARDS

Returning to the general water safety plan procedure, the final step is verification and, for the water utility, this will normally mean compliance with drinking water quality standards. The forms of numerical standards were covered in an earlier section. Europe, the United States of America, Japan and some other countries have developed their own approaches to the setting of standards; all of these countries take the WHO Guidelines into consideration but also take decisions on standards based on their own assessments. It is not sensible for other countries to consider making their own assessments, because the WHO guideline values themselves are based on the best scientific evidence available at the time, and it requires any other organisation to match that available expertise. Also, now that the WHO Guidelines are web-based with "continuous" updating, they provide the most up-to-date position on any parameter. As an alternative, should another country consider adopting European, American, or any other country's standards? I would say no because compared with the "purity" of WHO guidelines, nationally derived standards can be based on objectives other than healthy drinking water. Some standards have been set in the mistaken belief that drinking water standards can protect the environment. "Putting the horse after the cart" does not work. Drinking water standards should be based on health and aesthetic considerations, and the environment should be protected through a combination of environmental objectives and pollutant controls. The advice therefore is to adopt WHO Guidelines. Should a country have a concern over a substance not yet covered by WHO, it should produce its own interim standard based on available information, and then persuade WHO to take on that parameter for ongoing consideration and inclusion in the Guidelines.

In adopting WHO Guideline parameter values, a country should consider which parameters need to be included and the form and frequency of monitoring. In the European Directive [42] some of the parameters are considered to be mandatory and others are indicator parameters. The distinction is that those listed as mandatory must be complied with, whereas indicator parameters have to be monitored, and action taken only if an exceedance of the parameter value is a risk to health. This has the benefit of focussing investment on meeting those parameters of significance to health. The Directive also provides for two different monitoring regimes, those of check and audit monitoring. Check monitoring, with higher monitoring frequencies than audit monitoring, is required for microbiological, aesthetic and treatment parameters. Ideally, monitoring frequencies should be based on the expected variability over time. In practice it becomes too complicated to consider having an individual frequency for each parameter. It would also be very expensive, as it is much more cost-effective to use the same samples for a number of different analyses. The result is that sampling programmes have to be a compromise between that ideally needed to reflect the differences between parameters and practical considerations related to taking samples and cost.

The locations of sampling are important. It is not sufficient to sample at treatment works because the quality of water can be affected in a number of ways in the distribution system. The microbiological quality should be checked at the treatment works, at any storage points in the distribution, such as service reservoirs and water towers, and at consumers' taps. If there is an exceedance of a standard at any point, it is then possible to determine the source of contamination and make necessary improvements. Other parameters should be checked from samples taken from consumers' taps. One parameter is unique, at least in the way it is treated in the European Directive, and that is nitrite. There are two reasons for this. Firstly it needs to be considered in conjunction with nitrate. Secondly, it can be produced by microbiological activity in those distribution systems supplied by water which has been disinfected by chloramination rather than chlorination [43]. The health concern determining the EU approach to nitrate and nitrite combined, is the risk to bottle-fed babies of methemeoglobinemia (blue baby syndrome). The approach is to set a standard for nitrate of 50 mg jl at the tap, to set a standard for nitrite of 0.5 mg jl at the tap, but also to have a standard of 0.1 mg jl for nitrite at the treatment works. The combination of the two at the tap is controlled by the formula in which (nitrate/50+ nitrite/3) must be equal or less than 1. This formula is taken from the WHO Guidelines in which the guideline value for nitrite is 3 mg jl but, due to uncertainty on the impact of nitrite, the EU has opted for a much more stringent standard for nitrite. A review [44] of the guideline values for nitrate and nitrite was carried out by WHO in 2011. The guideline values have not changed. A brief description of each of the parameters in the EU Directive, as transposed into UK regulations, is given in Part 1 of the Drinking Water Inspectorate's Annual Report of 2005 [45].

It is not the purpose of this chapter to give a comprehensive review of parameters and standards but rather to identify the general issues to be considered in the establishment of standards and regulation. One such policy area is how to protect against *cryptosporidium*, with different approaches being taken in the USA and the UK. *Cryptosporidium parvum*, the form affecting humans and animals, is a single-celled animal (a protozoa), which can survive outside a host for months. The infective stage of the organism, the oocyst, is 3 microns in diameter, and millions of oocysts are excreted by infected humans and animals. They are always present in the environment. *Cryptosporidium* [46] is resistant to disinfection, particularly chlorine, so the best protection is to have processes which will physically remove the oocysts in treatment. In the United States, [47] the regulatory approach is to require specific levels of treatment, measured as numbers of log-removal (each log reduces the numbers of organisms to one tenth as described earlier in this chapter) according to the risks identified determined from two years of monitoring data. Each treatment process qualifies for so many log-removal. In England and Wales, following some serious outbreaks of waterborne cryptosporidiosis, the Government introduced specific regulations (1999) [48] involving a risk assessment, a numerical limit on the numbers of oocysts in a 24 hour continuously taken sample (with a minimum flow through the sampling device of 40 litres per hour), or the option of a treatment process capable of continuously removing particles of less than one micron in size. The limit of less than one oocyst in 10 litres of water was not based on a known infectious dose, but on experience in knowing that outbreaks of cryptosporidiosis have not occurred in "conventional" coagulation and filtration systems that are working normally. Exceedance of the standard was a criminal offence, but water companies had a defence of due diligence. The continuous sampling device was not only an important device for regulatory purposes, but also provides a very useful tool for water utilities to measure and optimise the performance of their treatment plant. The sampling result was much more accurate than so-called "grab" samples, providing for a much higher oocyst recovery. In 2006 following the virtual elimination of *cryptosporidium* incidents Government issued a consultation document, following which it was decided that continuous sampling monitoring was no longer necessary, and that public protection could be achieved through water safety plans. The changes [49] reduced the regulatory burden, which was a welcome step, and reflected the emphasis of a move to a risk management approach in line with the Bonn Charter and the WHO Drinking Water Guidelines.

INTERIM DRINKING WATER QUALITY STANDARDS

In countries where meeting standards based on WHO guideline values will require high investment over many years, it makes sense to set interim standards, with emphasis on meeting the microbiological standards. This has two benefits:

it provides the means of focussing investment over the next planning period, and also provides water utility managers with achievable targets. It can be very disheartening to know that the required standards are out of reach, and can result in a lack of motivation for improvement. On the other hand, achieving interim standards can give a sense of achievement and the motivation to achieve more. Once interim standards have been met the "bar can be raised" to aim for the next level of improved quality. Of course, setting interim standards requires care to ensure that parameters occurring at levels of immediate threat to health are included, with standards set which are achievable with available and affordable technology. It is important at the outset to have carried out an extensive survey of water source quality to identify which parameters should be included in the interim standards. It is not possible to be prescriptive on this, but in many cases it can be expected that there will be great benefit if the first level of interim standards are based wholly on microbiological considerations. Such interim standards should include chlorine residual measurements in the distribution system. However, interim standards should be a means to an end, and that end should be the eventual achievement of standards based on WHO guidelines.

A WORD ABOUT DISINFECTION

Disinfection, especially with the use of chlorine, is an emotive subject. The general public has an understandable concern over the use of a "dangerous" chemical- after all chlorine gas is used as a "chemical weapon". This concern is exacerbated by scare stories about disinfectant by-products being carcinogens. So what is the true situation? Yes, chlorine is a toxic gas, and great care has to be taken in its handling. The main risk is in handling it at a water treatment works, rather than in its use in disinfecting water. In fact, disinfection of drinking water with chlorine has probably saved more lives than all medication ever administered. There was a dramatic reduction in deaths from waterborne disease following the introduction of a combination of water filtration and disinfection in Europe and North America in the 19th century [50].

The concern over disinfection is due to the formation of undesirable substances which result from the reaction of chlorine with naturally occurring organic substances in the source water. Drinking water quality standards give limits to the levels of such substances in drinking water, and water treatment processes are designed to remove the organic content ahead of the chlorination process. However, very low levels of the undesirable chemicals remain. There have been extensive studies [51] over the last 30 years to try to measure any adverse impact on health. None has been conclusive, but if there is an effect, it is very small.

There are different views held by water consumers around the world about chlorine and its effect on the taste of water. In some areas of the United States of America, consumers feel uncomfortable if they cannot sense the presence of chlorine, because to them it is an indication of the safety of water. In some parts of Europe any sense of chlorine is unacceptable, and water utilities try to minimise the levels of chlorine residual in the distribution systems. In the Netherlands, in those systems that deliver only "dune" water, which contains no substances on which bacteria can grow in the distribution systems, there is no disinfection of the water supply. This is highly desirable, but unfortunately there are few distribution systems in the world where some means of controlling bacteriological contamination is not needed. Incidentally, any chlorine present in water arriving at consumers' taps will dissipate if left in a clean jug for a couple of hours. If kept in a refrigerator, the taste will also be improved, with people then being unable to distinguish between tap and bottled water in "blind tasting" tests. It is worth noting that the price of tap water is much lower, often only around one thousandth of that of bottled water. If people prefer to buy bottled water, and can afford to do so, then it is a matter of consumer choice, but in most developed countries, no one needs to buy bottled water for safety and health reasons. It must be our aim to extend that desirable situation throughout the world.

Whilst on the subject of aesthetic acceptability of tap water, a word about home-filter systems. Often they are used because people want to remove chlorine, or because they do not like "hard" water. The filters generally reduce hardness, which can mean removing desirable substances like calcium and magnesium [52]. The importance of magnesium is discussed in the next section. They can increase the level of sodium in the water which is undesirable and they can make the water highly acidic, so the advice to consumers is "make sure you know what the devices do to your tap water before using them". Remember the tip about putting a clean jug of tap water in a refrigerator for a couple of hours, to give a pleasant cool drink.

Returning to disinfection, there are alternatives to chlorine, the most common being ozone and ultra-violet light (uv), but neither of these provides ongoing control of bacteria in the distribution system. Ozone also produces disinfection by-products, and requires equivalent care in removing substances in the raw water to avoid their formation. The use of uv avoids such problems, but is not very effective if the water is "cloudy", having high turbidity.

Chlorination is likely to remain the most common disinfectant to achieve safe tap water in the developing world. Any concerns over by-products are tiny compared with the benefits in the reduction of waterborne disease. An important lesson [53] on this occurred in Peru in 1991–2. Following the publication of a USEPA report, there was concern over the formation of disinfection by-products, and the Government in Peru decided to stop disinfection of water supplies. In early 1991, there were the first reports of cholera, and by the end of 1992 there had been over 730,000 cases and 6323 deaths. The final death toll was estimated to be 11,000 deaths in the region. This was an over-reaction to an unproven possibility of a health effect from disinfectant by-products, which put a whole population at risk from a waterborne killer. This is why WHO advises that concern over chemical risks should never compromise the priority of protection from microbial pathogens through disinfection [18]. Also recently, WHO has published a report [24] on protecting drinking water in households with reference to the importance of disinfection.

THE IMPORTANCE OF MAGNESIUM TO HEALTH

WHO set up an expert team to review studies on hard water, calcium and magnesium to assess the evidence of beneficial effects regarding heart disease and strokes. The report [54] concluded that epidemiological evidence for the water hardness-cardiovascular hypothesis is still not proven. However, the balance of epidemiological evidence does support the (beneficial) link between magnesium and cardiovascular mortality. One of the studies [55] carried out in Taiwan over the period 1989–1993 showed a significant lower level of stroke deaths in those areas with higher magnesium levels in the drinking water. This understanding of the beneficial effects from magnesium in drinking water raises some important questions for regulation. Where minerals have to be added into water, for example in desalinated sea water or re-use water treated by reverse osmosis, should minimum magnesium levels be specified? Should devices, such as those home-filters which remove magnesium be regulated, or at least should there be warnings to householders that their use could be detrimental to health? My advice to householders is never to use water filters unless it is known what they do to the water. Suppliers have a moral responsibility to provide that information.

Whatever the form of water service provision, the private sector is always involved. Virtually all construction, whether it is for treatment works or underground pipe systems, is carried out by the private sector. The private sector provides virtually all of the hardware and consumables, and most of the software. Private sector consultants provide advice, develop schemes and carry out detailed design work. Private contractors carry out system maintenance, either in total or as a supplement to a water service utility's own workforce. In doing all this work, the private companies involved either make a profit, or they go out of business. In spite of a recognised profit element, it is accepted by those without bias, that the above mentioned work is more efficiently carried out by the private sector and that it results in benefits to water service customers. So why is the use of the private sector for financing and managing water services so controversial? Why do "failures" of private sector contracts receive extensive publicity and critical academic review? Why is it that poor performance in public sector operations, often in the same areas prior to the attempt to use the private sector, is generally ignored? This chapter attempts to look behind the emotive arguments, and to give advice based on an objective assessment of successes and failures. I begin with some examples of the history of the private water sector.

SOME EXAMPLES OF THE HISTORY OF THE PRIVATE WATER SECTOR

In London, there was private sector investment in water supplies as early as the 13th Century when French merchants, who wanted to be involved in the London wool markets, financed the Great Conduit [56]. This conduit, made of lead, conveyed water from what was the town of Tyburn, west of London, now Marble Arch, to the city area. Most people were then supplied by private water-bearers [57], who carried the water in 3 gallon or larger containers called tankards. The water collected from the conduit was free, but the water-bearers charged for their delivery service. There are various estimates of the number of water carriers, one as high as 4000, who formed an association known as the Fraternity to protect their interests. In later years, further conduits were built and from the

17th Century onwards, with the availability of steam power, there were a number of private initiatives to take water from the River Thames and deliver it through pipes to properties. The 30-mile long New River canal, taking water to London from Hertfordshire, was financed by a Welsh jeweller, Hugh Middleton. Over time, the responsibly for water supplies was taken over by the city and other local authorities, eventually all becoming operated by the Metropolitan Water Board, remaining public until privatisation of the Regional Water Authorities in 1989.

In the early days, permission to form a company to provide a water service required an act of parliament. Such permission was given to a man called William Yarnold in 1699 to supply water to Newcastle upon Tyne in the north of England. This company later became Newcastle and Gateshead Water Company, which was one of the 26 statutory private water companies included in the changes, when the Regional Water Authorities were privatised in England and Wales in 1989. Later, Newcastle And Gateshead Water Company became part of Northumbrian Water which exists today. Many of the statutory private water companies have a long history. These include Bristol Waterworks formed in 1846, Cambridge Water formed in 1853, Folkestone Waterworks Company (now Folkestone and Dover) formed in 1848, and South Staffordshire Water formed in 1853. Until 1989 these statutory water companies were regulated through a profit cap of 5%. So in England, water supply services owned and operated by the private sector are long established. They have a good reputation for integrity and service.

In France, there is also a long history [58] of the use of the private sector for the operation of water supplies. In the 18th Century private companies were granted concessions to take water from the River Seine and deliver the water in buckets to the citizens of Paris. In the 19th Century private companies began to build and operate water supply networks. The two large private companies, now known as Veolia and Suez, were formed around 1870. Today, about half of the water supplies in France are delivered by the private sector. The various forms of contract are discussed later in this chapter.

In The Netherlands, there is reference to Dune Water Company being formed in 1851 with English capital. Water was collected from the dunes and transported to Amsterdam by a large pipeline. In the book *Water of Vienna* [59] there is mention of "a number of private pipelines built in various parts of the city". It can be assumed that there were similar private initiatives throughout Europe during the early period of piped water supply.

The history in the United States of America is similar. One example is San Antonio [60] in Texas where in 1877 the city entered into a water supply contract with a private company. The company changed ownership a nmnber of times until 1925 when the city raised money through a bond issue and purchased the system.

THE EMOTIVE ARGUMENTS

There are probably more papers written either advocating the use of the private sector, or condemning it, that on any other water subject. The number of papers condemning it probably outnumbers those supporting it by ten to one. A highly organised anti-privatisation lobby tried to hijack the World Water Forum in

Kyoto in 2003. This had the unfortunate effect of attracting media attention, and overshadowing the more important issues on how to meet the water and sanitation needs of the unserved or underserved parts of the world. Given the history described above, why is the use of the private sector so controversial?

The main argument is that "water is a human right and should not be for profit". If that argmnent were to be taken to the ultimate degree, anything to do with water supply, including the manufacture of pipes and equipment, would have to be public sector. Equally, should the same argument apply to basic foods and basic shelter?

A more compelling argument is that since water supply is a monopoly, due to the high cost of the distribution infrastructure, it is not possible to have effective competition with the market controlling water pricing. It is largely true that direct competition is not possible, although there are elements of competition for larger customers using common carriage in the UK, and there are proposals to extend this to all non-household customers. However, an effective form of competition can be introduced through regulation-driven comparative competition.

There are also emotive arguments against the public sector, as some people believe that the public sector is inherently inefficient, as there are no market pressures to provide the motivation for management to strive for improvements. The World Bank was criticised ([61] and [62]) for adopting this attitude, generally by those opposed to privatisation. In my view, the Bank's policy needed to be challenged, but not for the reason it was being criticised. There was a need for the Bank to recognise that "one size doesn't fit all", that private sector involvement can provide substantial benefits in a well-regulated environment but that, in some other conditions, it is more appropriate to give emphasis to public utility reform. The Bank now puts a lot of effort into the capacity building of the public sector, although often this change of policy is not recognised by its critics.

I believe that all these emotive arguments are counter-productive, as there are enough examples to show that both the private sector and the public sector can be very successful in delivering good quality and cost-effective services. It is instructive to consider cases where the private sector has been criticised to help define what is necessary for its successful use.

PRINCIPLES OF EFFECTIVE PROCUREMENT

The problems experienced above can be attributed to a combination of lack of effective regulation and lack of effective procurement. In principle the approach to choosing a private sector company, to manage the operation of a water service, should be no different from selecting a contractor to build something, or an IT company to provide computers and information systems. However, there have been a number of highly publicised problems in the UK, such as the construction of Wembley Stadium, and provision of software for the Health Service, to illustrate that problems in procurement are not limited to the rather more emotive area of water service operations. Before considering sound procurement principles I list a number of different types of contract in the area of water services. Afterwards I focus my attention on the considerations necessary for successful concession contracts. The types of contract can be categorised as given in Table A.6.

Generally the principles for successful procurement apply equally to all types of contract. As effective procurement is a large subject, it can only be summarised in this book. For wider reading see [63] and [64]. A vital first step is for governments and utility managers to recognise that procurement and contract management are different from local government and direct operations management. Those without experience of procurement should seek expert advice. There is a need to avoid any pre-conceived views that private companies are by nature aliens, crooks or enemies. Yes, companies are in business to make money, but successful companies do that by providing good value to their clients in an honest way, without which they will not remain in business for long. Contractors should be encouraged to share the client's values and objectives, so that there are incentives for win-win situations with sharing of benefits. The

Table A.6 Types of contract in the construction and management of water services.

	Functions covered
Types of contract for whole systems	
Management	This involves only the operation of an existing system without any capital investment in either replacing existing assets or providing new ones
Affermage	This French term is used for a combination of operating and investment in the refurbishment of existing assets
Concession	This term is often used loosely to include also the two types of contract described above. A true concession contract involves the operation and refurbishment of existing assets and investment and operation of new assets
Types of contract for part systems	
DB (Design and Build)	This is limited to designing and building a new plant, or part of a plant, which on completion is handed over to others to operate
DBO (Design, Build and Operate)	Added to the above is the responsibility of operating the plant for a specified period after which it is handed over to the final owner, usually the public authority
BOT (Build, Operate and Transfer)	This is as a DBO except that the plant has been designed by others
Multi-systems	
Partnering	This can be applied to a single construction but generally involves a large programme of construction as in Scotland in the UK. It can also include both construction and operations as in the highly successful case of Wales in the UK. Both are described later in the chapter.

initial mindset is important in establishing a business climate of partnership, co-operation and openness. Much of that required for successful procurement is common sense. However, as there are so many failures, it has to be assumed that frequently the fundamental requirements for success are lacking.

In preparing a specification for contract bidding, there should be clarity of objectives, and clarity on what is included and what is excluded. Generally, outputs should be specified so that contractors can provide the benefit of their knowledge and experience. In the case of water treatment plant, specifying particular processes limits the scope of getting the best out of contractors. For this reason, I prefer DBO (Design, Build, Operate) to BOT (Build, Operate and Transfer) contracts. I prefer full concession contracts to either of these more narrowly-focussed contracts, and I return to that later.

It is important that sufficient time is taken to assemble the best information available, and to identify areas of significant uncertainty. Where there is uncertainty on the condition of assets, such as on the condition of the Atlanta distribution system described earlier, there is a need to build into the contract effort to obtain the necessary information, and to make provision for the contract terms to be adjusted accordingly at a later date. It is highly desirable that a draft specification is published. It provides transparency at the outset. It should also be presented in such a way that potential contractors are able to obtain clarification on requirements, and also to offer suggestions on improvement of the contract conditions. There should be sufficient time for this aspect, to ensure that there is clarity on the requirements of the contract, and that there is no "fudging" of the issues. There can be a temptation to "fudge", (that is to gloss over difficulties), in an attempt perhaps to meet some political deadline, but it is certain that eventually it will be necessary to address seriously overlooked problems. Inevitably, such contract deficiencies lead to higher costs, and have the potential for disputes between the parties and the potential for public criticism.

This raises the question of risk. Appropriate allocation of risk between the parties is key to success. There can be a tendency to require the contractor to take all the risks, but that approach does not make the risks go away. Either the contractors will include the accommodation of risk in the prices, or if a contractor is unable to live with the consequences, a contract will fail and the public authority will have to cope with those consequences. It is important that risks are allocated and borne by whichever party is best able to manage them.

Risk is associated with uncertainty and change. Contracts are generally for ten years or more, and it is important to recognise that circumstances will change over that period. Contracts need to be adaptable, with built-in procedures for addressing changing circumstances. It is better to have a partnership approach to change, with the parties working together to find the most cost-effective solutions. Contract "scoring" in which either party tries to win to the detriment of the other should be avoided, as conflicts which result in legal action only benefit lawyers. Should there be a major change in circumstances, there may be a need for immediate adjustment to the contract conditions. However, a number of small changes can be dealt with in a Periodic Review, perhaps every five years, when both requirements and contract payments can be adjusted in an integrated way. Such reviews are best overseen by an independent regulator, but more on that later. If there isn't an independent regulator, who can act as an arbitrator? Some other form of arbitration procedure should be agreed by both parties, and included in the contract. Should it be necessary to terminate the contract, perhaps due to a change in the politics of a country, or perhaps due to failure by the contractor to meet measurable outputs, there should be a termination procedure which is fair to both parties. In this respect, both compensation and penalty clauses may be included in the contract termination procedure. There are essentially three types of changing circumstances, which I consider in turn:

- Political/governmental;
- Legislative and currency changes;
- Physical aspects.

An example in the political/governmental category is a total change of attitude to the use of the private sector for water service operations. In Indonesia, some of the difficulties arose from the anti-Suharta backlash when the regime fell. In Bolivia, the change to a socialist Government has resulted in the termination of all concessions, whether or not they were working well. The same thing could have happened in Chile, but whilst the incoming socialist Government stopped the sale of the operating infrastructure to the private sector, it decided to continue with the use of the private sector through a concessionary approach. A reversal of policy can result from a change of mayor, as in Atlanta. Even changes of minister within the same government can trigger changes, as can reorganisation of government departments with a change of senior officials. The emergence of new anti or pro something bodies, most commonly anti-private pressure groups, can have a major impact on public opinion and thus government policies. These risks need to be assessed, particularly in countries with a history of political instability.

Changes in legislation can and do occur over the period of a contract. The most common ones are new drinking water quality standards, or new environmental standards, requiring increased investment. Generally, these can be predicted within a five-year cycle and provisions for the changes can be handled during the periodic review process. Handling such changes, which normally require an increase in water tariffs to pay for the additional investment, requires good communication with the general public, to ensure that they understand the reasons for the increases. If there is a communication vacuum, such changes provide the opportunity for anti-private groups to blame private sector "profits" for increases in tariffs made by governments. Other examples of legislation changes are changes to tax regimes and changes to general legislation impacting on procurement rules. Currency changes can be critical, especially if the cost of operation is influenced by another currency, such as the import of vital components or consumables. Where possible, every effort should be made to source locally.

Making provision for adjustments for physical changes is very important. The Atlanta case demonstrates how important it is to understand the condition of the systems, so that operating costs can be estimated with some accuracy. Where there is a lack of knowledge at the time of letting a contract, provision should be made to obtain the required information, and to adjust the contract conditions accordingly. A very common area of uncertainty is water distribution system leakage. It should be possible to know the level of non-revenue water fairly accurately, as it is the difference between the amount of water going into supply and the amount of water for which revenue is collected. However, in systems which have not been properly maintained, there are no reliable measurements of the amount of water leaving treatment works. This relies on having accurate meters at treatment works. This can be assessed fairly easily using insertion meters, but the proportion of non-revenue water, which is leakage, is more difficult to determine. It is vital that a full water balance [65] is carried out either before the contract is written or, failing that, to make it a condition of contract that it is carried out early in the contract. Reducing the level of non-revenue water is often one of the key objectives and performance measures, so having an accurate understanding of where all the water is going is essential. Another possible change in physical conditions is the quality of source waters, such as quality changes due to increased pollution. The provision to make contract adjustments for changes that are outside the control of the contractor should be included. Equally, should source water quality improve to the extent that the treatment costs reduce significantly, the reduced operational costs should be taken into account at a periodic review.

I have mentioned that the allocation of a risk in a contract should be based on which party is best able to manage it. The contractor should bear the risks associated with operations, and financial risks associated with construction projects, all of which should be under its control. Assuming that the contractor can operate without political interference, and has the ability to set up communications with the media and consumer groups, the contractor should be allocated risks associated with customer relations. This gives an incentive to provide a good service. If it is government that is letting the contract, whether national or local, it should undertake those risks associated with the impact of new regulations or tax changes, risks related to deterioration in source water quality, and risks associated with significant currency exchange rate changes. It is important that all these aspects are included explicitly in the contract to avoid subsequent arguments.

Establishing well-defined and well-drafted contracts requires knowledge, experience and sufficient time to work through all the issues. Extra time spent at the preparation stage can make the difference between success and failure, and avoid time, effort and cost later in dealing with contract deficiencies. It is important that, before inviting bids, the authority letting the contract should have a good estimate of the likely range of contract costs, and have well-defined selection procedures that take into account both cost and quality. The final contract specification should be checked carefully to ensure that all the issues have been addressed fully, and that no aspects have been "fudged". The need to avoid glossing over difficulties cannot be stressed too much.

PARTNERING

In recent years partnering has taken much greater prominence in the UK through the need to achieve very large construction programmes on time and to budget. Partnering is just that, partnering, in which two parties enter into an agreement to deliver something in an administratively efficient way, within a climate of trust, to avoid the conflicts that can arise from more traditional contracting, when the only beneficiaries are lawyers.

In partnering there are no invitations to tender for specific contracts. There are no conventional construction contracts. Instead, the customer and chosen suppliers enter into a partnering agreement, in which there is a contract framework with target prices. There is a so-called pam jgain approach through an open-book arrangement. If the contract outperforms, the partners share the gams. If there are problems and additional costs, the partners share the pain. Therefore there is a joint incentive to succeed. It relies on each partner understanding the other's goals, and sharing a commitment to achieve each other's goals to establish a win-win situation. It involves total transparency, sharing information and solving problems together. Instead of looking to blame each other, there is a culture of "how can we solve this problem together?" There has to be complete trust by each partner, with a demonstration of the required open behaviour to engender confidence in each other and in the partnering process. The approach is to start with facilitated workshops to establish the goals and the plans to achieve them, so the supplier is part of developing solutions right from the beginning. When problems arise, there are further workshops to find solutions to those problems. The management processes relate to managing the framework agreements, rather than monitoring the minutiae of individual contracts. The largest partnering agreement of which I am aware is for the delivery of major water and wastewater construction programmes in Scotland. There had been decades of under investment in water services in Scotland, resulting in the following circumstances in 2002:

- 78 wastewater treatment plants failed compliance with discharge standards;
- 40% of wastewater plants were considered "unfit for purpose";
- 60% of water mains and sewers were judged to be in poor or very poor condition;
- There were 9600 burst water mains every year.

Scottish Water, the corporatised-public water services company covering water and wastewater for the whole of the population of 5 million people in Scotland, estimated that £2.3 billion needed to be invested during the period 2002–6. The economic regulator, at that time the Water Commissioner, said that the Company did not need that amount and that it could achieve the required programme for £1.8 billion. As an indication of the scale of this lower figure, it amounted to around £360 for every person in Scotland. Scottish Water decided that the very large programme could not be delivered by conventional contracting, due to the sheer magnitude of letting such a large number of individual contracts in a short time period, and the need to complete the work at a total cost of 22% less than the estimated figure. The company therefore decided to use a partnering approach and formed Scottish Water Solutions. Scottish Water Solutions is 51% owned by Scottish Water, with consortia led by Thames Water and United Utilities owning 24.5% each. The consortia are made up of 8 companies, including large UK contractors. The Partnership was required to deliver 2300 projects in 2.5 years, amounting to 40% of all construction in Scotland during that period. By March 2004:

- 813 km of water main had been laid;
- 103 km of sewer had been replaced;
- Construction had been completed at 59 water treatment works;
- Construction had been completed at 20 wastewater treatment works;
- An additional inherited programme was 80% complete ahead of time and at lower cost;
- There were significant savings on the new construction programme.

One year later, the figures given in the 2004–5 Annual Report [66] were as follows:

- A £364 million investment programme had been delivered in that year;
- The regulatory output targets had been exceeded by 4%;
- 1176 km of water main had either been refurbished or replaced, setting a new UK record;
- Leakage had been reduced by 44 megalitresjday, equivalent to 6% of production;
- 156 km of sewers had been refurbished or replaced;
- 79 wastewater treatment plants had been either built or refurbished; and
- There had been improvements in drinking water quality and improved pressure in parts of the distribution systems.

In 2005–6, a £468 million investment programme was delivered, an increase of28% over 2004–5, with the following impressive statistics:

- Rehabilitation of 2774 km of water main;
- Rehabilitation of 337 km of sewers;
- Solutions to 322 unsatisfactory combined sewer overflows.

Scottish Water Improvements Report [67] gives information on what was delivered over the period 2006–2010. The capital programme which amounted to £260 per property per year over that period had peaked at £700 million in the year 2008/9. The £1.8 billion programme had largely been delivered on time. Some of the larger projects could not be completed in the time and some had been subject to planning delays. Most of the projects were to have beeen completed by the end of 2010. It is necessary to be a little cautious about that original figure of £2.3 billion, as it could have been part of the "regulatory game" of over-estimation. However, it is certain that the Regulator's allowance of £1.8 billion would have been very demanding. However, the report said that with experience of the capital programmes since 2002, investment should not exceed £500 million in any one year and that this would be taken into account in future programmes.

Scottish Water Solutions embarked on another major programme of partnering, this time with a different consortium and involving only above-ground asset construction. Water main and sewerage (underground) work would be handled internally. Although smaller than Scottish Water Solutions I, partly because the underground work was excluded, the partnering contract [68] involved 280 projects costing £450 million over the period 2010–2015.

I include this example, not simply to give an illustration of successful partnering on a large scale, but also to show how the public and private sectors can work together in a noncontroversial way. Scottish Water Solutions is an example of an effective public-private partnership.

A PERFECT CONCESSION CONTRACT?

Utopia in outsourcing water service operations and development might be considered to be the "perfect" concession contract. Can such a perfect contract exist? Perhaps not, but it is possible to let very good contracts which have an excellent chance of avoiding the problems experienced in the examples discussed in this chapter. I see the following key ingredients:

- A concession of a size large enough to attract good management and to give economies of scale, which probably means a minimum of 250,000 connections. This need not be within one hydraulic system, but could be the management and operation of a number of systems;
- A contract period of around 15 years. This is long enough to achieve management commitment and some continuity, but short enough to provide for the potential benefit of re-tendering. Contracts involving a high level of investment by the contractor may need to be longer, to alleviate the impact on charges by providing for cost recovery over a long period of time;
- Five-year periodic reviews;
- Clarity of requirements with clear objectives, sound information on condition of assets and reliable revenue forecasts;
- Good consumer consultation beforehand with a requirement for the contractor to handle consumer relations;
- The contractor given full responsibility to manage with no outside interference;
- An independent regulator to oversee the letting of the contract and to oversee the periodic reviews. Failing this, the naming of an agreed arbitrator in the contract;
- Tariffs based on need only, and not increased to fund other local services; and, above all;
- Transparency, with contract conditions published with good information provided to consumers. This is particularly important if the introduction of a contract coincides with increased tariffs in a move to sustainable cost recovery.

REGULATION IN CONCESSION CONTRACT SITUATIONS

In Chapter 2, I discussed "regulation by contract" saying that a well-defined contract provides the means of controlling the delivery of the desired performance of a water service operator, whether public or private. I said that it is a major step forward, and much better than the "traditional" municipal operation in which there are limited or no performance measures, and poor financial ring fencing. But I stressed that it does not provide any means of regulating the contract owner, nor does it provide for transparency and public participation. I went on to say that, on the contrary, it tends to reduce transparency, as it assumes that governments can look after the consumers' interests. This lack of transparency risks the creation of the conditions for corruption. So some form of independent oversight is required. This could involve regulating the contract owner. Dŵr Cymru is the "public" licensed "water undertaker" providing water services to Wales, and is regulated in the same way as the private companies in England. During the period when its operations were contracted out it continued to be the regulated utility. It had to take full responsibility for any failures by its contractors. Although the Dŵr Cymru situation is rather unique and the economic regulator was not involved in the contracts, the deliverables were defined by the regulatory system (See periodic planning in Chapter 3). More generally, without such a rigorous regulated planning system, an independent regulator can ensure that the conditions, on which a contract is to be based, are clear, objective and realistic. I believe that effective independent regulation could have avoided many of the problems described in the examples given earlier in this chapter. I do not see regulators interfering with the contracts, but providing the objectivity and transparency on which the contracts are based.

REFERENCES

1 Regulation of the Private Provision of Public-related Water Services. Economic Commission for Latin America and the Caribbean, 1997.

2 Groom E., Halpern J. and Ehrhardt D. (2006). Explanatory Notes on Key Topics in the Regulation of Water and Sanitation Services. Water Supply and Sanitation Sector Board Discussion Paper Series, Paper No. 6, World Bank, June 2006.

3 UNDP Human Development Report (2006). Beyond Scarcity: Power, Poverty and the Global Water Crisis. ISBN: 9780230500587.

4 Selmick P. (1985). Focussing organisation research on regulation. In: Regulation Policy and Social Sciences, R. Noll (ed.), University of California Press, Berkeley, California.

5 Tenenbaum B. (1995). The Real World of Power Sector Regulation. Viewpoint Note No 50. The World Bank.

6 Baldwin R. and Cave M. (1999). Understanding Regulation: Theory and Practice. Oxford University Press, Oxford, UK.

7 Mcintosh A. C. (2003). Water Supplies: Reaching the Urban Poor. Asian Development Bank. August 2003. ISBN: 9789715613804.

8 Saghir J. (2006). Keynote Address at International Water Association Congress, Beijing.

9 Eijlander P. (2005). Possibilities and constraints in the use of self-regulation and coregulation in legislative policy. *Electronic Journal of Comparative Law*, **9**(1), http://www.ejcl.org/91/abs91-1.html (accessed July 2013)

10 Ontario Water Strategy Expert Panel (2005). Watertight: The Case for Change in Ontario's Water and Wastewater Sector. Report of the Water Strategy Expert Panel. Queen's Printer for Ontario, 2005. ISBN: 9780779479283.

11 Personal Communication.

12 Johnston S. (2006). The Ghost Map. Riverhead Books. Published by the Penguin Group, ISBN: 9781594482694.
13 Mortality and Burden of Disease from Water and Sanitation (2004). Global Health Observatory. WHO, 2004.
14 Parasites- Giardia (2012). Centers for Disease Control and Prevention. July 2012.
15 Kappus K. D., Lundgren R. G., Jr., Juranek D. D., Roberts J. M. and Spencer H. C. (1994). Intestinal parasitism in the United States: update on a continuing problem. *Am. J. Trap. Med. Hyg.*, **50**(6), 705–713.
16 Global Burden of Disease Study (2010). GBDlO. www.healthmetricsandevaluation.org (accessed July 2013)
17 The Bonn Charter for Safe Drinking Water (2004). Published by the International Water Association.
18 Guidelines for Drinking Water Quality (2004). World Health Organization 2004. ISBN: 9789241546386.
19 Hutton G. and Haller L. (2004). Costs and Benefits of Water and Sanitation Improvements at the Global Level (Evaluation of the). World Health Organization 2004. WHOSDEWSH04.04.
20 Water Safety Plans: Managing Drinking Water Quality from Catchment to Consumer (2005). WHOJSDEJWSH/ 05.06. World Health Organization 2005.
21 Water Safety Portal. WHO and International Water Association. www.wsportal.org (accessed July 2013)
22 Water Safety Planning for Small Community Water Supplies: Step by step risk management guidance for drinking water supplies in small communities. (2012). World Health Organisation 2012, ISBN: 9789241548427.
23 Hazard and Risk. (2013). Index of Educational Resources. Centre for Occupational and Environmental Health, UK. http://www.medicine.manchester.ac.uk/oeh/undergraduate/onlineresources/hazardandrisk/ (accessed July 2013)
24 Combating waterborne disease at the household level. WHO, ISBN: 9789241595223.
25 Toolkit for Monitoring and Evaluating Household Water Treatment and Safe Storage Programmes. (2012). WHO/ UNICEF 2012. ISBN: 9789241504621.
26 Preventing Cryptosporidiosis: the Need for Safe Drinking Water (2013). World Health Organisation, 2013. 2013-91:238, doi: www.who.int/bulletin/volumes/91/4/13-119990/ (accessed July 2013)
27 Breach R. and Williams T. (2006). The Pivotal Role of Water Safety Plans, Bonn Principle Series. IWA 2006.
28 Blue Drop Requirements (2011). Department of Water Affairs. Republic of South Africa. http://www.dwaf.gov.za /dir_ws/DWQR/ (accessed July 2013)
29 Hodgson K. and Manus L. (2006). A drinking water quality framework for South Africa: ISSN 0378-4738. Water SA, **32**(5). http:/ /www .ajol.info/index.php/wsa / article/view/47853 (accessed July 2013)
30 Policy Issues and Institutional Arrangements: Water Safety Plans for Rural Water Supply in India. Water and Sanitation Programme Report. November 2010. http://wsp.org/sites/wsp.org/files/publications/Water-Safety-Plans-rural-water-supply-India.pdf (accessed July 2013)
31 Drinking Water Safety: Guidance to Water and Health Professionals. (2009). Published by Drinking Water Inspectorate for England and Wales. ISBN: 9781905852406.
32 The Scottish Water (Objectives for 1 April 2010 to 31 March 2015) Directions 2009. www.scotland.gov.uk /Resource/Doc/ 917 /0088613.pdf (accessed July 2013)
33 Water and Health in Hungary: Report of Second Meeting of the Parties to the Protocol on Water and Health. Joint Secretariat to Protocol on Water and Health. http://www.unece.org/env/water/whmop2.html (accessed July 2013)

34 Drinking Water Safety Plannning: A Practical Guide for Pacific Island Countries. WHO/SOPAC Joint Contribution Report. http://www.pacificwater.org/userfiles/file/water%20publication/Pacific%20Drinking%20Water%20Safety%20Planning%20Guidelines%20WHO%20SOPAC.pdf (accessed July 2013)
35 Mullenger J., Ryan G. and Hearn J. (2002). A Water Authority's experience with HACCP. Water Supply, **2**(5–6), 149–155. IWA Publishing 2002.
36 Miller R., Whitehill B. and Deere D. (2005). A national approach to risk assessment for drinking water catchments in Australia. *Water Supply*, **5**(2), 123–134.
37 Australian Drinking Water Guidelines (2006). Ref. EH19, EH39. National Health and Medical Research Council Publications, Australia.
38 Stevens M. (2006). Water Safety Plans in Australia. Presentation at WHO Workshop, IWA Congress, Beijing, September 2006.
39 Godfrey S., Nivagaba C., Howard G. and Tibatemwa S. Water Safety Plans for Developing Countries. A Case Study from Kampala. Uganda. Watermark website: http://wedc.lboro.ac.uk/resources/conference/30/Tibatemwa.pdf (accessed July 2013)
40 Examples of water safety plans in rural areas in Bangladesh. Website: http://www.buet. ac.bd/itn/ pages /apsudocs /WSP_full_document.pdf (accessed July 2013)
41 Water Tanker Service Guidelines for the Utility and Operators of Tanker Services. (2007). Public Utilities Regulatory Commission, Ghana 2007.
42 Drinking Water Directive (98/83/EC) http://eur-lex.europa.eu /LexUriServ/LexUriServ.do?uri=OJ:L:1998:330:0032:0054:EN:PDF (accessed July 2013)
43 Regan J., Harrington G. and Noguera D. (2002). Ammonia and nitrite-oxidising bacterial communities in a pilot-scale chloraruinated drinking water distribution system. Applied and Environmental Microbiology, January 2002, 73–81.
44 Nitrate and nitrite in drinking water: Background document for development of WHO Guidelines for Drinking-water Quality. (2011). WHO/SDE/WSH/07.01/16/Rev/l. World Health Organisation 2011.
45 Drinking Water (2005). A Report by the Chief Inspector, Drinking Water Inspectorate. London, June 2006.
46 The Third Report of the Expert Group on Cryptosporidium in Water Supplies. (1998). Drinking Water Inspectorate, London.
47 Long-Term 2 Enhanced Surface Water Treatment Rule. (2006). US Environmental Protection Agency, Washington DC, January 2006.
48 The Water Supply (Water Quality) (Amendment) Regulations (1999). Drinking Water 2000. A Report by the Chief Inspector, Drinking Water Inspectorate. London, June 2001.
49 The Water Supply (Water Quality) (England and Wales) Regulations (2000). SI No.3184 (Unofficial Consolidated version 2010) (PDF 896KB) Website: http:// dwi. defra.gov. uk/stakeholders/legislation jwqregs2007cons.pdf (accessed July 2013)
50 Governance for Sustainable Human Development (1997). A UNDP Policy Document. United Nations Development Programme, January 1997.
51 O'Neill G., Mole N., Horth H. and Fielding M. (1994). Disinfectant By-products: Occurrence and Significance to Health. Report No FR0481 Foundation for Water Research, November 1994.
52 Report on WHO Meeting of Experts on the Possible Protective Effect of Hard Water Against Cardiovascular Disease, held in Washington DC, USA April 2006. Public Health and Environment WHO Geneva 2006.
53 Chlorination M. X. and the Law of Unintended Consequences (August 2003). Kimballs Biology. http://users.rcn.com/jkimball.ma.ultranet/BiologyPages/C/Chlorination. html (accessed July 2013)
54 Calcium and Magnesium in Drinking Water: Public Health Significance (2009). World Health Organisation 2009. ISBN: 9789241563550.

55 Yang Chun-Yuh Calcium and Magnesium in Drinking Water and Risk of Death From Cerebrovascular Disease. http://stroke.ahajoumals.org/content/29/2/411.full (accessed July 2013)
56 Trench R. and Hillman E. (1984). London under London. John Murray, London, ISBN: 9780719552885.
57 Flaxman T. and Jackson T. (2004). Sweet & Wholesome Water. Flaxman & Jackson, England, ISBN: 9780954898601.
58 Payen G. (2004). Local Management and Competition. IWA Workshop, Tokyo.
59 Egghardt H., Lammerhuber L. and Riha G. (2003). The Water of Vienna. Bohmann. ISBN: 9783901983221.
60 San Antonio Water System History and Chronology. www.saws.org (accessed July 2013)
61 Bayliss K. and Hall D. (2002). Unsustainable conditions- the World Bank, privatisation, water and energy. August 2002. www.psiru.org (accessed July 2013)
62 Bank 'soul searching' on privatization. April 2003. www.brettonwoodsproject.org (accessed July 2013)
63 Erridge A., Mcilroy J. and Fee R. (2001). Best Practice Procurement. Gower Publishing Ltd Aldershot, UK.
64 The Orange Book (2004). Management of Risk-Principles and Concepts. HM Treasury, UK.
65 Farley M. and Trow S. (2001). Losses in Water Distribution Networks: A Practitioner's Guide to Assessment, Monitoring and Control. IWA Publishing. ISBN:9781900222112.
66 Rising to the Challenge. Scottish Water Solutions. Annual Report and Accounts, 2004–5. www.scottishwater.co.uk (accessed July 2013)
67 Investing in Scotland's Water Industry: Improvements Delivered in 2006–10. Scottish Government. www.scotland.gov.uk/Resource/Doc /917/O/112271.pdf (accessed July 2013)
68 Scottish Water Solutions 2: Building on Success (2010). Presentation at Institute of Water Autumn Seminar, 17 September 2010, Institute of Water, Tyne and Wear, UK.